Palgrave Studies in Alternative Education

This series emerges out of a recent global rise of interest in and actual educational practices done with voice, choice, freedoms and interpersonal thoughtfulness. From subversion to introversion, including alternative settings of the state to alternative pathways of the private, the series embraces a diverse range of voices.

Common to books in the series is a vision of education already in existence and knowledge of education possible here and now. Theoretical ideas with potential to be enacted or influential in lived practice are also a part of what we offer with the books.

This series repositions what we deem as valuable educationally by accepting the power of many different forces such as silence, love, joy, despair, confusion, curiosity, failure, attachments as all potentially viable, interesting, useful elements in educational stories. Nothing is rejected if it has history or record as being of worth to people educationally, nor does this series doubt or distrust compelling ideas of difference as relevant.

We wish to allow mainstream and marginal practices to meet here without prejudice as Other but also with a view to ensuring platforms for the Other to find community and understanding with others.

The following are the primary aims of the series:

- To publish new work on education with a distinctive voice.
- To enable alternative education to find a mainstream profile.
- To publish research that draws with interdisciplinary expertise on pertinent materials for interpersonal change or adjustments of approach towards greater voice.
- To show education as without borders or boundaries placed on what is possible to think and do.

If you would like to submit a proposal or discuss a project in more detail please contact: Rebecca Wyde rebecca.wyde@palgrave.com. The series will include both monographs and edited collections and Palgrave Pivot formats.

More information about this series at
http://www.palgrave.com/gp/series/15489

Elizabeth A. C. Rushton

Science Education and Teacher Professional Development

Combining Learning with Research

palgrave
macmillan

Elizabeth A. C. Rushton
King's College London
London, UK

Palgrave Studies in Alternative Education
ISBN 978-3-030-64106-1 ISBN 978-3-030-64107-8 (eBook)
https://doi.org/10.1007/978-3-030-64107-8

This Palgrave Macmillan imprint is published by the registered company Springer Nature Switzerland AG.
The registered company address is: Gewerbestrasse 11, 6330 Cham, Switzerland

This book is dedicated to David, Matthew, Nathaniel and Lucas who, together, provide the compass points to my life.

Foreword

Over 30 years ago, I went into teaching because of my love of physics and science in general and my hope that I could help others see the excitement, power and potential in science. As a science teacher, this passion needs to exist but is regularly squashed by the need to deliver the required specification and get students to then flourish in their exams. Somewhere along the way, the essence of what we are trying to do as teachers often gets lost in the need to maximise results and improve attainment. I, of course, want students to succeed but this cannot be at the expense of making science some rigid set of notes with limited opportunities for practical work. Students will say that a subject is interesting when the teachers are interested and excited by it. For science teachers that means being involved in what is going on in science right now. The idea that science teachers stop being scientists when they chose teaching limits them hugely.

I first met Lizzie when we both were teaching at the same secondary school in south-east England. We could see that science was more interesting for all when teachers, technicians and students were involved at the cutting edge. We took all opportunities offered and that created a 'can-do' atmosphere for students and staff. As a teacher, this was the science education I wanted for my students and, of course, I found it fascinating and inspirational too. Over time, more teachers wanted this experience of research-led teaching, being a part of the scientific effort alongside university students and academics.

This book comes at an opportune moment in science education to set the scene for a better approach to professional development opportunities for teachers and technicians. It is the result of years of research that Lizzie has conducted in schools and with the Institute for Research in Schools (IRIS). IRIS developed to spearhead the spread of this vision of an authentic science education for teachers and students across the country. The 'teacher scientist' model of professional identity, articulated so clearly by Lizzie here, sums up this shift to a better experience for the teacher and the students and a more fulfilling experience of a science education. Lizzie and I worked together on many projects within IRIS, and Lizzie's rigour and quest for academically watertight evidence led to her phenomenal work interviewing teachers and technicians and finding out what their experiences were. This book brings together a wealth of evidence to argue persuasively for this approach to science education and professional development. Lizzie's sharp analysis shows that the experience of teachers and technicians as 'teacher scientists' is overwhelmingly positive and should feed into approaches needed now to tackle persistent science teacher recruitment and retention challenges.

I hope it will be transformative in showing a powerful new model for science teacher professional development and, I am sure, it will inspire many across the fields of education and science to consider new approaches and policy initiatives concerning science teacher education.

Kent, UK
London, UK
Summer 2020

Becky Parker

Preface

As with many books, the ideas shared in this one are rooted in my own experiences. My career in teaching began as a doctoral researcher when, in my 'writing up' year at the end of three years of funding, and with two small sons to support, I found myself paid employment as a teaching assistant in a local secondary school. During this time, my days were filled trying to support and engage young people in varied subjects including geography, maths and science. Long evenings were spent trying to pull together a thesis about socio-ecological interactions in Belize, Central America, spanning a 3500-year period which explored Maya, colonial Spanish and British and present-day land-use using documentary and environmental records. Somehow, I managed to complete my PhD and subsequently trained as a secondary school geography teacher, and because my work as a teacher began almost literally submerged in my research, my identity as a researcher, as a geographer, did not leave me when I entered the classroom. In turn, my experiences as a teacher strengthened my belief in the value of the discipline of geography, of research and of the capacity of young people to meaningfully contribute if given the opportunity.

As such, whilst working as a secondary school geography teacher, I ran research projects with sixth form students (aged 16–18 years) as a part of an extra-curricular programme where the only pre-requisites were enthusiasm, interest and the willingness to give up a Wednesday afternoon each week. Projects included analysis of samples taken from ancient woodland in Oxfordshire to explore climate change and using the archival records of a local Cathedral to consider peoples' changing understandings of climate over time. My role as a researcher and a teacher was to provide students with the skills (e.g. laboratory techniques and locating and analysing documents) and access to networks and information necessary for them to explore what the data might mean, to pose questions and to think of ways to share their interests and understandings with others. Students who joined these projects were usually not identified as 'high attainers'; they were a wonderful collection of people interested in the world around them and who frequently found that they had something to contribute and to offer but always had no space or context to share their ideas and perspectives. I had the great fortune to watch these young people as they engaged in research projects each week for at least a year at a time. Over this period, they developed the knowledge, skills, networks and confidence to ask their own questions and search for the answers and were frequently able to persuade others that what they had learnt and done was of value.

It is a huge privilege that Prof Becky Parker has agreed to write the foreword for this book. I met Becky when I was a trainee teacher, and my ideas about how geography teachers and students could engage with and contribute to research were given flight by her boundless energy and enthusiasm for the ways in which young people can and do contribute to scientific research. Becky has championed this cause for decades and her ongoing legacy are the hundreds of teachers and thousands of students whose education has been enriched by working as part of the many science projects that she has developed and encouraged. Working with Becky as part of the Institute for Research in School gave me the opportunity to meet and work with science teachers and technicians across the UK and beyond, who have been and continue to be research active with their students.

Through interviews with these teachers and technicians shared in this book, I have sought to understand how working in research can and does provide an alternative approach to teaching secondary school science that benefits both students and their teachers and technicians. Understanding the ways in which participation in research projects can provide an enjoyable and beneficial form of professional development for teachers and technicians is not something that would seem 'alternative' in a university setting. Indeed, in higher education, there is the expectation that teaching and learning is 'research-led' or 'research-informed' and that lecturers bring insights from their own research activities to their students as part of their day-to-day practice. However, this is still quite an alternative conceptualisation of school teacher professional development where teachers are enacting and implementing curricula that are largely determined by others and do not usually have the opportunity to develop their subject knowledge through being research active in their various disciplines. This book does not argue that participation in research projects is the only or even a superior form of professional development. However, I hope I make a case that involvement in research projects is an alternative approach that deserves careful consideration by school leaders and policy makers if we are to recruit and retain teachers who will continue to inspire the students they teach.

London, UK
July 2020

Elizabeth A. C. Rushton

Acknowledgements

The impetus for this work came from my time working with school students, teachers and technicians who, supported by the Institute for Research in Schools (IRIS), contribute to a range of STEM-focused research projects. I am grateful to all those who shared their time, experiences and expertise, and I am indebted to the 53 teachers and technicians who generously agreed to be interviewed as part of this research.

I received much encouragement and support from the staff and trustees of IRIS, especially Steve Greenwood, Mike Grocott and Laura Thomas. During my time as a secondary school teacher and subsequently at IRIS, I have had the good fortune to work with and learn from Prof Becky Parker (Founder of IRIS). Becky's enthusiasm and generosity has been and continues to be invaluable. Thank you also to Prof Michael Reiss (UCL Institute of Education) and Prof Helen Walkington (Oxford Brookes University) for supporting my research and for their encouragement to write this book.

I am grateful to the team at Palgrave for carefully guiding this novice author through the process of writing and producing a book, especially Dr Helen Lees, Prof Michael Reiss, Rebecca Wyde and the two reviewers, who provided helpful feedback on the book proposal.

Since joining the School of Education, Communication and Society, King's College London, I have benefited hugely from the kind, supportive and active research community of The Centre for Research in

Education in Science, Technology, Engineering & Mathematics (CRESTEM). I would particularly like to thank Dr Heather King for her encouragement and good humour.

I am very grateful to my father, Dr David Millard, for reading through drafts of this book and providing wise and detailed feedback. Lastly, I would like to thank my husband Matthew, and our sons, Nathaniel and Lucas, for being the most supportive and loving 'home team' anyone could ask for.

Praise for *Science Education and Teacher Professional Development: Combining Learning with Research*

"By exploring the positioning of Science educators as producers rather than consumers of authentic educational research, this book provides a compelling and radical rethink of the oft-cited term *evidence-based teaching.* It offers new insights of how becoming research-active shapes Science educators' professional identity and their senses of self."

—Prof Lulu Healy, *Professor of Mathematics Education, King's College London, UK*

"A valuable and timely resource, this book will be essential reading for teachers who want to connect—or remain connected—with scientific research and to inspire the young people they teach through independent research projects. This book makes an important contribution to our understanding of science teacher identity."

—Dr Lynda Dunlop, *Senior Lecturer in Science Education, University of York, UK*

Contents

Abbreviations

ACP	Alternative Certification Programmes
A-level	Advanced Level—exams taken by students at the end of their high school education
CPD	Continuing professional development
DSMRI	The Dynamic Systems Model of Role Identity
GCSE	General Certificate of Secondary Education—exams taken by students at the end of the fifth year of high school education
GRS	Get Real! Science
EPQ	Extended Project Qualification—usually taken by students in their final year of high school and is equivalent to half an Advanced Level qualification
IRP	Independent Research Project
IRIS	Institute for Research in Schools
ISI	Informal Science Institutions
ITT	Initial Teacher Training/Education
MELT	Monitoring the Environment, Learning for Tomorrow
NSLN	National STEM Learning Network
PD	Professional Development
PL	Professional Learning
PLC	Professional Learning Community
QTS	Qualified Teacher Status
RCT	Randomised Controlled Trials
RTA	Reflexive Thematic Analysis

SCP	The Scientist in the Classroom Partnership
SIA	Social Identity Approach
SKE	Subject Knowledge Enhancement
STAI	State-Trait Anxiety Index
STEM	Science, Technology, Engineering and Mathematics
STI	The Science Teacher Identity Model
TA	Thematic Analysis

List of Tables

1

Science Education and Teacher Professional Development

1.1 Introduction

This chapter considers the challenging context of teacher recruitment and retention, which is particularly acute for science teachers, in the UK and beyond, and explores different ways to recruit and retain teachers including financial incentives, alternative certification programmes and subject-specific continuing professional development programmes. Key concepts from professional learning are also considered, including teacher professional development and the place of mentoring and coaching in the educational context. This chapter begins by briefly outlining the impacts that high teacher turnover has on students and teachers and situates this within the wider context of young people's equitable participation in STEM education and careers.

1.1.1 STEM Education and Careers

It is widely recognised that high teacher turnover has a negative impact on both teachers and students (Allen & Sims, 2018). Attainment is lower where there is high teacher turnover (Atteberry, Loeb, & Wyckoff, 2016;

E. A. C. Rushton, *Science Education and Teacher Professional Development*, Palgrave Studies in Alternative Education, https://doi.org/10.1007/978-3-030-64107-8_1

Ronfeldt, Loeb, & Wyckoff, 2012) and the impacts of high turnover are exacerbated in shortage subjects such as science as in order to fill vacancies, school leaders (particularly those working in disadvantaged contexts) frequently have to lower their recruitment standards, rely on temporary cover and/or increase class sizes (Allen & Sims, 2017; Smithers & Robinson, 2000). As Allen and Sims (2017) highlight, these compensatory strategies have themselves all been linked to a decrease in student attainment (Fredriksson, Öckert, & Oosterbeek, 2013; Mocetti, 2012). All teachers become more effective during their first three to five years in the profession (Papay & Kraft, 2015) and particularly science teachers (Henry, Fortner, & Bastian, 2012), and the replacement of those who leave the profession entirely with newly qualified teachers further affects student attainment. The impact on teachers who leave the profession, especially those who leave in the first five years, is also highly negative, with those individuals often reporting feelings of professional failure and low self-worth (Allen & Sims, 2018). Difficulties in recruitment and retention of science teachers that are experienced to varying degrees across Europe and the USA are also mirrored in the low numbers of young people progressing to STEM courses in higher education and careers, even in wealthy countries promoting equal opportunities and STEM careers (e.g. UK, France; OECD, 2015), and, paradoxically, also where national gender equality gaps are lowest (e.g. Finland, Norway; Stoet & Geary, 2018). This reflects persistent inequalities in the accessibility of scientific career options and deprives STEM research of the full diversity of human capital needed to advance these fields. Previous research has revealed various factors influencing students' aspirations and progression to STEM careers, especially their general science knowledge, and beliefs and attitudes towards studying science (Archer et al., 2013), for example, valuing science and scientists, self-efficacy beliefs (Archer, Dawson, DeWitt, Seakins, & Wong, 2015), experiences in science (Rushton, Charters, & Reiss, 2019) and own imagined future careers (DeWitt & Archer, 2015; Sheldrake, Mujtaba, & Reiss, 2017a). Studies show the critical role of teachers in shaping students' intended STEM course enrolment (Mujtaba & Reiss, 2013; Sheldrake, Mujtaba, & Reiss, 2017b) and the potential of student-teacher STEM research collaborations for strengthening school teachers' own identity with science and scientists (Rushton & Reiss, 2019).

The concept of 'science capital'—a way of understanding an individual's science-related social and cultural capital—conceived by Archer et al. (2015) has been developed to articulate the ways in which an individual's science-related knowledge, understanding and experiences are valued, in certain settings. Science capital extends from work of the French sociologist Pierre Bourdieu (1977) who theorised the notions of *capital*—the social, cultural and economic resources an individual may have; *habitus*—the dispositions an individual may hold and *field*—the physical setting and social relationships which frame the ways in which one's capital and dispositions are valued, or not. This concept provides a lens through which to explore why some children are able to negotiate and benefit from science learning opportunities, both in 'the fields' of the classroom and out-of-school settings. The concept of science capital has been developed into a teaching approach that can support more equitable engagement (Godec, King, & Archer, 2017). The foundation of the Science Capital Teaching Approach (SCTA) is to (1) broaden what counts as science, (2) personalise and localise young people's encounters with science, (3) elicit, value and link the contributions young people make and (4) incorporate science capital dimensions (Godec et al., 2017). A science capital lens to young people's engagement in science argues that if we are to acknowledge and value young people's varied funds of knowledge and make explicit the links between their everyday and school-based knowledge, traditional learning conventions and structures should be reconsidered. The SCTA has focused on outcomes for student participation in post-16 years STEM courses and careers, rather than the recruitment and retention of science teachers. Similarly, initiatives that support student engagement in school-based science research have primarily considered the benefits for students, rather than the implications for teachers (Bennett, Dunlop, Knox, Reiss, & Torrance Jenkins, 2018; Eldridge, Lock, & Vokins, 1995; Reiss, 1992; Stock Jones, Annable, Billingham, & MacDonald, 2016). Research has highlighted links between this school science learning grounded in research and inquiry and increased science attainment (Furtak, Seidel, Iverson, & Briggs, 2012; Mehalik, Doppelt, & Schunn, 2008; Minner, Levy, & Century, 2010), more positive attitudes towards STEM and STEM careers (Gibson & Chase, 2002; Ornstein, 2006) and increase participation in STEM subjects (Rushton & Parker, 2019).

This book presents a new perspective to research that considers the impact of school-based science research because it focuses on outcomes for teachers. A central argument is that teachers' participation in school-based research provides an alternative for sustained and dynamic subject-specific continuing professional development (CPD) by (re)connecting teachers with their identity as a scientist.

This book challenges the implicit assumption that it is only teachers of certain subjects, for example, music, art, drama and English, who can enhance their classroom teaching with the continued practice of their craft as an orchestral instrumentalist, landscape painter, actor or poet. Before advancing this argument in detail, the broader challenges of secondary school teacher recruitment and retention are considered.

1.1.2 Secondary Teacher Recruitment and Retention

Teacher recruitment and retention is a global if not universal challenge: an OECD (Organisation for Economic Co-operation and Development) PISA (Programme for International Student Assessment) survey of education systems in 69 countries found that 29% of 15-year-olds were learning in schools where school leaders reported that a lack of teaching staff had limited the school's capacity to provide instruction (OECD, 2018). Countries which continue to experience shortages include England (Foster, 2018; Sibieta, 2018), Scotland, Wales and Northern Ireland (Davies et al., 2016) and a number of other wealthy countries including Ireland (O'Doherty & Hartford, 2018), the USA (García & Weiss, 2019), Australia (Weldon, 2018), as well as parts of continental Europe (Eurydice, 2018) for example Sweden (Skolverket, 2017). The numbers of qualified teachers leaving the profession has also become problematic, with retention especially difficult in the first five years of teaching. The figures for full-time equivalent teachers across both primary and secondary settings in England show an increasing 'wastage rate', rising from 9% in 2011 to 10% in 2017 and, in 2017, more teachers left the profession than joined it (Foster, 2018). In England, attrition rates are highest in the first five years of a teaching, with 20% leaving in the first two years, rising to 33% within the first five years (DfE, 2018). Similarly, in the USA, attrition rates in 2015/2016 are at 8%, compared to values of 3–4% in countries including Finland, Singapore and Canada, and only one-third of US teachers who leave the profession return (Sutcher, Darling-Hammond, & Carver-Thomas, 2016). Although Weldon (2018) suggests that teacher attrition rates are challenging to measure in the Australian context, researchers and media reports indicate that 30–50% of teachers leave the profession within the first five years, and perhaps 25% of these return to the profession at a later date. Schools in socio-economically disadvantaged contexts are more likely to face both recruitment and retention problems (NAO, 2016, 2017) although, drawing on the research of Johnson, Kraft, and Papay (2012), Allen and Sims (2018) highlight that teachers are not leaving these environments because of the disadvantage they encounter, but that they are less likely to receive the support they require in the early years of their career. As teacher recruitment and retention rates decline, pupil numbers are increasing, exacerbating an already problematic issue for school leaders and policy makers. In England, secondary school populations are projected to increase by 15% in 2025 compared to 2018 (Foster, 2018). In the USA, the number of teachers needed is estimated to rise from 64,000 in 2015/2016 to 316,000 in 2025 (Sutcher et al., 2016). These increases are due to both growing student populations as well as a government-led initiative to reduce teacher-pupil ratios (Sutcher et al., 2016).

1.1.3 Secondary Science Teacher Recruitment and Retention

Recruitment to secondary Initial Teacher Education (ITE) courses in England during 2019/2020 was 15% below government targets, with significant shortages in chemistry, computer science, mathematics and physics. The only science subject to meet, and indeed exceed its recruitment target, was biology (DfE, 2019b). Since 2001, the Norwegian government has recognised the challenges its education system faces in recruiting science teachers (Forni, 2007), even in a country acknowledged to have extremely high quality of life and well-respected education system (HDI, 2006). Over the past 15 years, policies to address teacher attrition, including increases in salary, have been successful; however, recruitment strategies have failed to increase the numbers of science graduates entering postgraduate teaching training due to the diversity of alternative highly paid careers (Forni, 2007).

As well as recruitment challenges, England has had a substantial shortage of qualified science teachers, sustained since 2013 (MAC, 2016; Sims, 2017). Only 50% of physics teachers remain in the profession five years post-qualification (Sibieta, 2018) and there is an increased likelihood of science specialists leaving the profession (Worth, De Lazzari, & Hillary, 2017). Allen and Sims (2017) found that attrition rates for science teachers in the first five years are 26% higher than colleagues from other subjects. Attrition rates for science teachers with a physics/engineering degree are 29% higher than for non-science teachers (Allen & Sims, 2017). As with general secondary teacher shortages in England, rising pupil numbers (19% increase in secondary-age pupils by 2026) will worsen the impacts of under-recruitment and retention of science teachers, particularly as entries for triple science GCSE are increasing (Sims, 2019). Science teacher recruitment and retention in England is therefore, particularly, challenging in subjects including chemistry, physics and mathematics, as well as computer science (NAO, 2016), and these shortages are replicated across the UK and in other wealthy countries, including America, Australia and continental Europe (Rushton & Reiss, 2020). Over the past five years in Sweden, schools have had unfilled secondary teacher vacancies in subjects including science and mathematics which have resulted in government policies to encourage well-qualified science graduates to change careers and train to be teachers (Molander & Hamza, 2018; Skolverket, 2017).

The USA also has high and sustained shortages of science teachers; during 2015/2016, 40 states had insufficient science teachers to fill existing vacancies (Sutcher et al., 2016). Since 2000, over 10% of US schools have reported challenges in recruiting enough science teachers, and Sutcher et al. (2016) suggest that this is caused by an underproduction of science teachers rather than high attrition rates, as science graduates have varied opportunities for employment with much higher levels of compensation. Sutcher et al. (2016) suggest that as demand rapidly increases, alongside increasing pupil populations, subjects (such as science) with current recruitment challenges are likely to face more severe shortages in the future.

In summary, over the past two decades, across economically well-resourced countries, there have been high attrition rates of qualified science teachers and under-recruitment to postgraduate teacher training courses. These trends have resulted in a range of government policies to address these challenges, including alternative training programmes, salary enhancements, subject-specific continuing professional development (CPD) and professional partnerships with scientists. Before exploring a range of policy solutions to the persistent challenge of science teacher recruitment and retention, the range of terms found in the literature that are associated with in-service teacher education are reviewed, including: professional development, continuing professional development, professional learning, professional learning communities, coaching and mentoring. These terms are briefly described and defined at the outset of discussions about science teacher recruitment and retention that immediately follow and are also relevant terms to consider ahead of the review of science teacher identity and professional development; that is presented in Chap. 2.

1.2 Teacher Professional Development and Learning

A substantial body of research suggests that improving the effectiveness of teachers is the single most important factor within schools that policy makers can influence to improve student attainment (Machin & Murphy, 2011; Rivkin, Hanushek, & Kain, 2005; Rockoff, 2004). High-quality teacher instruction is consistently positively associated with pupil outcomes (Hattie, 2009) and school improvement (Jackson & Temperley, 2007). Drawing on Gusky (2002), Ufnar and Shepherd (2019) define teacher professional development (PD) or continuing professional development (CPD) as 'the process and activities used to enhance professional knowledge and skills of teachers, and to change attitudes toward teaching to increase student achievement' (p. 642). Research has suggested that effective PD has a focus on content, active and sustained learning and coherence with other learning opportunities (Darling-Hammond, Hyler, & Gardner, 2017; Darling-Hammond & Richardson, 2009; Ufnar & Shepherd, 2019), as well as being implemented as part of a well-thought out plan that includes teacher feedback throughout (Mizell, 2010). Darling-Hammond et al. (2017) define effective teacher PD as 'structured professional learning that results in changes to teacher knowledge and practices, and improvements in student learning outcomes' (p. 2). Effective PD has clearly established benefits for teachers, pupils and the wider school community (Yoon, Duncan, Lee, Scarloss, & Shapley, 2007) and subject-specific PD has been linked to better pupil outcomes than generic pedagogic PD (Cordingley et al., 2018).

Teachers' professional learning (PL) is developed through both externally provided courses and 'job-embedded' activities that shape their practices and thus, PD is one part of a range of experiences that may result in professional learning (Darling-Hammond et al., 2017, p. 2). Professional Learning Communities (PLCs) are groups of teachers from different schools who work with educators and other related subject specialists with the aim of enhancing professional learning and school improvement (Binkhorst, Handelzalts, Poortman, & Van Joolingen, 2015; Schaap et al., 2019). PLCs feature productive discussions of teaching (via face-to-face meetings or electronic communication) and can provide teachers with a source of support that extends beyond their school environment. PLCs are often thematic (e.g. increasing attainment for students with Special Educational Needs) or subject-specific (e.g. physics, geography) and teachers often receive financial and/or practical support as part of their engagement (Schaap et al., 2019). Productive and supportive PLCs are places where expertise is shared through networks that are sustained overtime (Bell & Cordingley, 2014); however, their impact is acknowledged as variable (Kennedy, 2016) and Schaap et al. (2019) document the tensions that teachers experience during their participation in PLCs that are frequently linked to workload and a lack of shared learning.

As part of PLCs, teachers are frequently supported by mentors or coaches who may be based in university departments, either as subject or pedagogical specialists. There are substantial bodies of literature that explore the roles of both mentors and coaches within teacher education and professional learning (Ambrosetti & Dekkers, 2010; Lofthouse, 2019). Teacher education programmes include mentoring, which occurs during school-based professional placements where trainee teachers are mentored by classroom teachers to learn and develop their teaching knowledge and skills (Ambrosetti & Dekkers, 2010). In a review of the literature, Ambrosetti and Dekkers (2010) acknowledge that mentoring is a dynamic, complex and multifaceted role that is contextually based and involves both process aspects (e.g. developing learning opportunities) and relationship aspects (e.g. providing emotional support) (Ambrosetti & Dekkers, 2010). Mentoring is also a feature of teachers' continuing professional developing and learning, with Lofthouse (2018) arguing that mentoring is a 'dynamic hub' for both individual professional learning and institutional growth that allows mentors, mentees and teacher educators to contribute to the transformation of professional learning practices. A related role in teacher professional learning is that of a coach: coaching is increasingly a feature of teacher education and leadership programmes, although as Lofthouse (2019) notes, there is a lack of understanding as to the efficacy of coaching and no recognised or validated qualifications for coaching in educational contexts. As with mentoring, coaching is understood variously depending on the cultural context of the PD practice (Lofthouse, 2019) and because coaching approaches themselves are shaped by the particular circumstances of the school, teacher and coaching programme (Kennedy, 2014). Coaching has been positively linked to teacher professional development (Veenman & Denessen, 2001), teacher self-efficacy (Tschannen-Moran & Tschannen-Moran, 2010) and student outcomes (Joyce & Showers, 1988); however, Lofthouse and Leat (2013) note that the pervasive focus on performance in schools can undermine the potential of coaching.

These understandings of terms, including and relating to teacher professional development, provide a useful context to the following discussion of science teacher recruitment and retention; this considers a range of approaches, including financial incentives, alternative training programmes, subject-specific continuing professional development and engagement with science research and teacher professional development.

1.3 Recent Approaches to Recruit and Retain Science Teachers

In a systematic review of science teacher identity, Rushton and Reiss (2020) identified three broad themes found in studies from the UK, the USA and Sweden that explore recruitment and science teacher education in the context of alternative training programmes: (1) a broad consideration of the role of alternative training programmes; (2) the role of informal science institutions in science teacher training and (3) training programmes with reform-minded and/or social justice approaches. Other recruitment approaches include paying trainee-teachers tax-free bursaries during their training period and salary enhancements in order to retain early-career teachers (Sims, 2017). Research that considers science teacher identity highlights the importance of a teachers' science/subject identity (Rushton & Reiss, 2020). Similarly, some retention strategies for qualified science teachers include participation in professional development programmes including those with a focus on teacher engagement with research (Rushton et al., 2019; Rushton & Reiss, 2019) and classroom-based partnerships with scientists (Ufnar, Bolger, & Shepherd, 2017; Ufnar & Shepherd, 2019).

1.3.1 Financial Incentives to Recruit and Retain Teachers

In an attempt to ameliorate teacher shortages, the English government, working with professional bodies, have introduced bursaries and competitive scholarships for those individuals training in shortage subjects. These financial awards are made at various scales depending on their subject, degree classification and the high-needs status of their school context. For example in England, high school teacher trainee entrants may be eligible for bursaries ranging between £6000 and £26,000 depending on the subject, and competitive scholarships of £28,000 available in physics, chemistry, modern foreign languages and computer science (Foster, 2019). Post-qualification, the take-home pay of trainee teachers who are awarded the highest scholarships can actually decrease and, as Sims (2017) highlights, bursaries and scholarships do not address the retention problem. Research has demonstrated that the causes of teacher shortages are varied and often context-specific, variously linked to high workload, challenging working conditions, geographical remoteness and the high cost of living (O'Doherty & Hartford, 2018). Pay is particularly important in regions and countries where teachers' salaries have not kept pace with other employment opportunities (e.g. USA, Allegretto & Mischel, 2016). Although not universal across Europe (exceptions include Ireland and Finland), the teaching profession is widely considered to have reduced prestige, with this decrease linked to comparatively low pay and the deterioration in working conditions (O'Doherty & Hartford, 2018).

Rates of teacher pay are relevant to the retention of science and mathematics teachers, as STEM graduates are likely to receive higher rates of pay in careers outside of teaching, making salaries an important factor (Sims, 2017). Recent research from the USA has shown that increasing science teacher salaries has a large positive effect on retention, at least in the short-term (Bueno & Sass, 2016). Drawing on empirical evidence from North Carolina and Florida, Sibieta (2018) highlights how salary supplements for mathematics and science teachers worth approximately 5% reduced teacher wastage rate by between 10 and 20% and that bonus payments targeted to attract high-ability teachers to high-needs areas have been an effective strategy in US states including California. In 2017, the English government introduced two new policies relating to the remuneration of early-career, shortage-subject teachers. Firstly, science and modern foreign language teachers working in priority areas would be eligible to have their student loan repaid during the first decade of their career and secondly, mathematics teachers training during 2018–2019 will receive a supplement of up to £7500 in their third and fifth years of teaching if they work consistently in state-funded schools in England post-qualification (Sims, 2017). In 2019, these early-career payments were increased to £9000 for those working in areas of high deprivation and would be paid to physics and chemistry graduates, as well as those with degrees in modern foreign languages (DfE, 2019a). Modelling by Sims (2017) suggests that had a 5% salary supplement for science and mathematics teachers been introduced in England during 2010–2015, shortages of teachers in both subjects could have been eliminated.

1.3.2 Alternative Training Programmes

In England, about 30,000 people enter ITE courses each year. There are several routes available; however, the key distinction is whether training is predominantly based in school (e.g. School Direct, *Teach First*) where a trainee often receives a salary or whether they are based in higher education, and pay tuition fees (Foster, 2019). As has been described, there is significant variation in the funding available to trainee teachers, with those training in shortage subjects including mathematics, science and geography eligible for tax-free bursaries and scholarships. Regardless of training route or financial package, all trainees are required to have experience in at least two schools and on successful completion are awarded Qualified Teacher Status (QTS). Entrance requirements for all ITE courses included GCSEs at a minimum of grade C in English, mathematics and, for primary ITE, science is also a pre-requisite. Across Europe, there is the broad requirement for teachers to have a bachelor's degree and a teaching qualification (European Commission, 2018); however, in some contexts it is mandatory to have a postgraduate qualification (e.g. in Scotland and Finland) whilst other countries do not require a separate postgraduate qualification to obtain Qualified Teacher Status (e.g. England, Sweden) (O'Doherty & Hartford, 2018). In the USA, the traditional teaching training route is through a four-year undergraduate degree, although, as in England and Sweden, there is a greater diversity of training routes. These different routes are collectively described as alternative certification programmes (ACPs) and are usually a shorter course of study (e.g. 5 weeks–18 months) that is completed by individuals who already hold an undergraduate degree in a science or related subject.

In their review of the literature, Rushton and Reiss (2020) identified that research that considers middle and/or high school science teacher alternative training programmes broadly fall into three themes: (1) a broad consideration of the role of alternative training programmes; (2) the role of informal science institutions in science teacher training and (3) training programmes that were explicitly and predominantly framed around reform-minded and/or social justice approaches. Preservice teachers training in the alternative routes described in the US-based research (e.g. Friedrichsen, Lannin, Abell, Arbaugh, & Volkmann, 2008; Proweller & Mitchener, 2004; West, 2015) enrolled in programmes that were specifically designed to address the lack of specialist mathematics and science teachers that is particularly acutely felt in economically poor areas of rural (West, 2015) and urban regions (Proweller & Mitchener, 2004) of the USA. West (2015) highlighted that the shortage of science teachers in the USA has led every state to introduce an ACP, and perhaps the most well-known programme is *Teach for America*, established in 1990, where graduates complete a five-week summer school and gain the necessary state licences required to teach as a *Teach for America Teacher*, working in high-needs contexts. A similar programme called *Teach First* has been established in England and Wales for over 15 years, training in excess of 10,000 teachers to work in low-income areas (Teach First, 2019). Friedrichsen et al. (2008) suggested that in the USA, the introduction of ACPs has led to an increase in the diversity of trainee teachers, who have a greater range of non-teaching experiences. Woolhouse and Cochrane (2015) explored the experiences of teachers who complete Subject Knowledge Enhancement courses (SKEs). These courses usually last 8–28 weeks and are undertaken by those training to teach physics, chemistry or mathematics who have a pre-existing specialism in a different but related area (e.g. biology, psychology). SKEs enable those with a closely related degree or relevant professional subject experience to meet the subject requirements for ITE. In England, SKEs are available in the sciences, mathematics, modern foreign languages and geography and there are some tax-free bursaries available. In recent times, teacher training routes have been developed specifically for PhD graduates, with the aim to encourage highly qualified STEM specialists into school classrooms. These include the three-year *Researchers in Schools* programme run by the

England-based educational charity, *The Brilliant Club* (The Brilliant Club, 2020) and a nine-month programme funded and implemented by the Swedish government since 2017 (Molander & Hamza, 2018).

As well as a higher education institutions and education charities, informal science institutions (ISIs) have also provided spaces for trainee science teacher education. Adams and Gupta (2017) have considered the role of ISIs and suggest that they enable teachers to develop teacher identities that include student-centred approaches where teachers are responsive to the needs of learners. Adams and Gupta (2017) also suggested that these spaces support teachers to imagine their future professional selves and classrooms that maintain these identities. Heredia and Yu (2017) also found that ISIs provide novice high school science teachers with subject-specific support and expertise that was rooted in a community of practice of 'like-minded' science teachers. Subject-specific guidance was recognised by McIntyre and Hobson (2016) as an important source of identity development for beginning teachers of physics based in the UK; however, in this research, subject-specific support was provided by external mentors rather than ISIs. McIntyre and Hobson (2016) suggested that external mentors (experienced subject-specialist teachers working at different schools to their mentees) provide a space where mentees could discuss their professional learning and development needs and could, with the mentors' support, identify alternatives to performative norms.

Studies that have considered alternative teaching training includes research that is explicitly set within a framework of social justice (Luehmann, 2007; Luehmann, 2016; Richmond, 2016; Rivera Maulucci, 2013; Rivera Maulucci & Fann, 2016) or reform-minded approaches (Danielowich, 2012). Luehmann (2016) described reform-based science teaching for social justice as a pedagogy which actively works towards social equity, in the context of more accurate and critical understandings of science for all students, especially those already less well-served by the education sector. Rivera Maulucci (2013) defined social justice teaching approaches as 'an ongoing struggle for more caring, equitable, and agentic schooling at classroom (micro), school (meso), and community/society (macro) levels' (p. 454). Developing as a socially just teacher is not without difficulty or challenge, and researchers in this field have clearly highlighted the emotional ambivalence that individuals face relating to their personal and professional identities (Richmond, 2016; Rivera Maulucci, 2013; Rivera Maulucci & Fann, 2016). In an exploration of a ten-year science education programme called *Get Real! Science* (GRS), which focused on developing science teachers who are committed to social justice, Luehmann (2016) highlighted the importance of out-of-school contexts (e.g. after-school science clubs) that provide low-risk settings to explore professional identity development. The need for these non-judgemental spaces, where preservice teachers can explore and reflect upon their emerging identity, is also commented on by McIntyre and Hobson (2016). These 'third spaces' may be situated in physically different settings (e.g. ISIs) or manifest through alternative professional relationships (e.g. with external mentors). What is common across the use of these spaces is a lack of judgement and the professional freedom to explore and experiment (Adams & Gupta, 2017; Luehmann, 2016; McIntyre & Hobson, 2016).

Luehmann (2007) highlighted the emotional challenges that reform-minded teachers face and the need for periods of reflection in order to undertake 'repair work' necessary to continue developing this identity. Other research that has considered the place of identity and emotions in the development of social justice and reform-minded teachers contributes frameworks which can inform the training and continuing professional development of educators (Richmond, 2016; Rivera Maulucci, 2013; Rivera Maulucci & Fann, 2016). In the context of sustained science teacher shortage in high-poverty schools in the USA, Richmond (2016) argued that developing teachers' sense of commitment to working in these challenging contexts is crucial for successful recruitment and retention of individuals appropriate for this setting.

Another approach to science teacher recruitment is to attract scientists or those with a science-background to move from their current careers into teaching. Research that explores the experiences of 'career changers' is now considered.

1.3.3 Teachers Who Are Career Changers

Studies that have considered the experiences of teachers who have transitioned into teaching from a previous career in a STEM field (as opposed to a postgraduate research degree, such as a PhD) include two that focused on qualified science teachers working in the USA (Antink-Meyer & Brown, 2017; Snyder, Oliveira, & Paska, 2013), and three that explored the experiences of career changers during a one-year training course in the USA (Grier & Johnston, 2009), Australia (Watters & Diezmann, 2015) and the UK (Wilson & Deaney, 2010). Understanding the experiences of science teachers who are career changers is regarded as important in these studies due to an increasing number of recruitment strategies and training schemes which directly target career changes, in an attempt to address the international shortage of science teachers. Broadly speaking, the focus of the five studies is around the transition of professional identities from those rooted in a STEM career to that of a science teacher and the particular challenges and opportunities this transition represents to the individuals concerned as well as to ITE providers. In seeking to theorise this transition, the five studies draw on a range of theories and frameworks including identity (Antink-Meyer & Brown, 2017; Grier & Johnston, 2009; Molander & Hamza, 2018); self-determination theory (SDT) (Watters & Diezmann, 2015) and transformative learning theory (TLT) (Snyder et al., 2013). Rushton and Reiss (2020) offered two distinct but related commentaries on these studies: (1) there is a recognition across all five studies that career changers benefit from ITE that is the same as or broadly similar to ITE aimed at those following more traditional postgraduate routes; and (2) career changers frequently have multiple sources of identity on which to draw; these studies tend to present these changes as transitions into and out of different identity states (e.g. scientist, parent, teacher), rather than understanding these transitions as individuals developing a range of group memberships.

A consistent, but perhaps unsurprising, finding across these five studies that considered career changes is the importance of teacher education courses as a site of intersection between identities of scientist and science teacher. Snyder et al. (2013) suggested that the transition into teaching from a STEM career leads to the emergence of a new sense of self that is not always compatible with previous STEM identities and, during this transition, ITE courses can provide important spaces that support individuals to reconcile different aspects of identity. Grier and Johnston (2009, p. 73) go further and argued that ITE programmes are an 'essential' part of the process of identity formation of science teachers from any route or background. Part of the acknowledged value of ITE is the opportunity for explicit teaching on identity formation to support novice teachers in developing an awareness of their changing sense of identity (Wilson & Deaney, 2010). Researchers have documented how ITE can acknowledge and leverage STEM career contexts when supporting individuals through this period of transition and provide opportunities for career changers to become reflective practitioners (Grier & Johnston, 2009), although Wilson and Deaney (2010) suggested that teaching on identity formation should be embedded in ITE. Not all studies have found that ITE courses were effective in meeting the particular needs of mature career changers training to be teachers. Watters and Diezmann (2015) argued that ITE courses need to be more adept at recognising and valuing the prior experience of this group of trainee teachers. Watters and Diezmann (2015) also highlighted the importance of the school context for career changers as a site of professional identity formation. It should be noted that this body of research on career changers emanates from researchers who are also intrinsically linked to developing and delivering teacher education programmes, some of which may include their research participants (Snyder et al., 2013; Wilson & Deaney, 2010). Although research that includes student teachers as participants is not uncommon, it should be acknowledged that studies that are based on very small participant numbers and are drawn from the researcher's own context may limit the replicability of findings across broader contexts.

Having considered a range of recruitment strategies including the payment of tax-free bursaries, and alternative training programmes we explore a variety of initiatives that seek to retain qualified science teachers within the profession.

1.3.4 Subject-Specific Professional Development

Researchers have explored how a science teachers' identity is shaped by both their identity as a teacher and their science/subject identity and have found that for both qualified and preservice teachers, their identities are most closely aligned with their science/subject identity. For example, in a study that analysed survey responses from 80 high school teachers, 27 of whom taught science and mathematics, Beijaard, Verloop, and Vermunt (2000) suggested that teachers saw themselves in terms of their subject rather than pedagogical experts and that this was especially true for teachers at the beginning of their career, for science and mathematics teachers and for male teachers. This finding of the importance of science/subject identity over teaching identity in a study from the Netherlands that is nearly 20 years old is consistent with more recent research from a range of locations including the UK (Irving-Bell, 2018; Woolhouse & Cochrane, 2015), Chile (Ortega, Correa Molina, & Fuentealba Jara, 2014), the USA (Chung-Parsons & Bailey, 2019; El Nagdi, Leammukda, & Roehrig, 2018) and Canada (Nieswandt, Barrett, & McEneaney, 2013). Participants in this diverse group of studies were more likely to define themselves in relation to a shared identity around their subject, and therefore more likely to enact the norms and values associated with that identity than those of a teaching identity. Relatedly, Irving-Bell (2018) found that subject knowledge was a core part of establishing positive professional identities. Woolhouse and Cochrane (2015) explored identity developing in trainee science teachers who participated in a Subject Knowledge Enhancement (SKE) course in physics, chemistry or mathematics, having a pre-existing specialism in a different but related area (e.g. biology, psychology) and again found that their subject identity was clearly developed at the beginning of their teaching careers. Manning (2017) found that a teacher or pedagogical identity grew over time, as in a study of urban high school science teachers who had been teaching for an average of five years, these teachers' identity was more strongly aligned with that of an urban school teacher as opposed to a science teacher. Rushton and Reiss (2020) suggest that teachers at the beginning of their careers have not yet had time to form identities that are shaped by the practice of teaching, and therefore, they have identities that are predominantly rooted in their subject.

The importance of subject-related identities for science teachers is particularly relevant when considering the place of subject-specific continuing professional development programmes in retaining science teachers. In identifying eight principles for increasing the quantity and quality of science teachers, Sims (2019) suggested that science teachers should be able to increase their specialisation so that they have the opportunity to focus their teaching in one science subject, preferably the subject they are qualified to degree level and that this is especially important for early-career teachers as they develop their teaching practice. Sims (2019) also highlighted the importance of science-specific professional development, as he argues that this approach to continuing professional development is linked to increased science teacher retention (Allen & Sims, 2017), due to increased efficacy and job satisfaction (Sims, 2019; Wolstenholme, Coldwell, & Stevens, 2012). Allen and Sims (2017) explored the impact of participation in England's National STEM Learning Network (NSLN) professional development courses during 2010/2011 and 2012/2013. During this period, 83% of all secondary schools in England had at least one teacher attend an NSLN course and 57% of secondary schools had teachers attend at least five days' worth of courses over this three-year period (Allen & Sims, 2017). The impact evaluation revealed that teachers who participated in an NSLN course are around 160% more likely to be retained in the profession in the year after they participated in the professional development course, and that this association is still visible two years after participation for both experienced and more recently qualified teachers (Allen & Sims, 2017). When Allen and Sims (2017) analysed the impact of participation at a department rather than an individual level, they observed a 4% reduction in the proportion of science teachers leaving teaching in the two years after at least one of the department's teachers participates in an NSLN course. As the average attrition rate for English secondary science departments during this period is 10% per year, a reduction by 4% decreases the attrition rate to 6% which therefore makes a substantial contribution to science teacher retention (Allen & Sims, 2017). Subject-specific professional development courses support science teacher retention, and this may in part be linked to the importance of a teachers' subject/science identity, especially early in their career. Another approach to science teacher professional development provides teachers with opportunities to engage with science research, and in the following section examples from the UK (Rushton & Reiss, 2019), the USA (Varelas, House, & Wenzel, 2005) and Norway (Mehli & Bungum, 2013) are briefly explored.

1.3.5 Teacher Engagement with Science Research

Studies that considered the experiences of science teachers who engage in professional development programmes that are rooted in research-based approaches include Mehli and Bungum (2013), who identified that science teachers belong to several communities of practice including teacher, science teacher and scientist, and suggest that the focus of the literature on teachers' professional development is on their role as educators rather than their identity as scientists. This is perhaps unsurprising, as research has demonstrated that science teachers, particularly preservice and early-career teachers, strongly identify with their subject (e.g. Beijaard et al., 2000; Manning, 2017), and developing an identity as an educator is crucial to be an effective teacher. However, Mehli and Bungum (2013) argued that professional development that is situated in authentic science research contexts gives teachers an opportunity to engage and work with science professionals in a way that provides insights into scientific practice in modern society. Mehli and Bungum (2013) outlined how these experiences were formed as part of a week-long residential science camp and suggested that they were beneficial when developing science teachers' subject-specific professional identities. This is consistent with the findings of Rushton and Reiss (2019), who considered the professional identity development of 17 high school science teachers. This group of teachers (with their students) collaborated over at least a six-month period in authentic science research with scientists, who were based in universities and research institutes. Rushton and Reiss (2019) described how these teachers, at varying career stages and with a range of prior experience of science research, developed multifaceted professional identities that enabled them to be both science teachers and 'teacher scientists'. Rushton and Reiss (2019) argued that the key enabling factor in this identity development is the teachers' positive interaction with scientists and/or researchers.

There are key differences in the genesis and implementation of the professional development programmes considered in the research of Rushton and Reiss (2019) and Mehli and Bungum (2013). For example, in Rushton and Reiss (2019), teachers worked with their students, predominantly in their school environments with the engagement of scientists and/or researchers over an extended period of time (at least six months). In contrast, in the research of Mehli and Bungum (2013), teachers, gathered from across Norway, participated in a week-long residential course located at a space technology research site. However, even though the model of delivery is distinct, the outcomes are broadly similar: teachers in both programmes developed their identities as both teachers and scientists through engagement with 'real scientists' and 'authentic science' in a way that enhanced their professional identity beyond the initial life of the programme. Varelas et al. (2005) considered the identity development of four beginning science teachers who completed a ten-week summer apprenticeship in a science laboratory. Teachers in this study identified features of science as a practice (e.g. theory and data, complexity) and a community of practice (collaboration, autonomy, mentoring) and were positive about the freedom and the community of working in the laboratory space. However, some teachers expressed tension and conflict when bringing their experiences of the apprenticeship to their identities as science teachers and described difficulty in incorporating their scientist identity into their classroom (Varelas et al., 2005). The research of Varelas et al. (2005) demonstrated the importance of supportive professional environments (in this case schools, departments and classrooms) and networks (colleagues and senior leaders) if teachers are to successfully transfer nascent facets of their identities from one context to another. A related example of professional development that involves teacher interaction with scientists and current science is a US-based programme called 'the Scientist in the Classroom Partnership' (Ufnar et al., 2017; Ufnar, Kuner, & Shepherd, 2012; Ufnar & Shepherd, 2019).

Scientist in the Classroom Partnership

The Scientist in the Classroom Partnership (SCP) programme places a STEM graduate student in a classroom one day per week to co-teach with the classroom teacher over the course of one academic year (Ufnar et al., 2012; Ufnar et al., 2017; Ufnar & Shepherd, 2019). The original purpose of the programme was to help STEM graduate students enhance their communication and teaching skills rather than outcomes focused on school students or teacher professional development; however, Ufnar et al. (2017) found that how fellows and teachers participated in the programme reported increase in confidence, use of inquiry in the classroom and pedagogical skills. Ufnar and Shepherd (2019) suggested that the SCP programme is a replicable model that exhibits characteristics of effective teacher professional development. Drawing on teacher responses to a survey and contributions to focus groups, Ufnar and Shepherd (2019) explored the experiences of over 50 of the 74 teachers who participated in the SCP programme during 2000–2009. This identified five core and three structural features. The structural features include a focus on (1) use of multiple types of professional development approaches—including complementary delivery approaches including summer workshops, job-embedded co-teaching and planning sessions, (2) the duration of the professional development—professional development is more likely to be effective if it is substantial and sustained over a long period of time and the SCP programme consisted of 200 hours over an academic year and (3) teachers working with the same school or year group (Ufnar & Shepherd, 2019). Core features of the professional development model include (1) gains in discipline content knowledge, (2) gains in pedagogical knowledge, (3) learning and implementing inquiry and hands-on strategies, (4) collaboration within and outside the classroom and (5) teacher renewal (Ufnar & Shepherd, 2019).

The SCP is another example of the different approaches to teacher professional development, many of which are frequently linked to teacher recruitment and retention, and include developing subject-specific expertise as well as partnerships with educators and research-active scientists. In Chap. 2, these ideas of the different relationships and contexts are considered as part of a wider understanding of the formation and reformation of science teacher identity and the factors that influence and shape this over time.

References

Adams, J. D., & Gupta, P. (2017). Informal science institutions and learning to teach: An examination of identity, agency, and affordances. *Journal of Research in Science Teaching, 54*(1), 121–138.

Allegretto, S. A., & Mischel, L. (2016). *The teacher pay gap is wider than ever*. Washington, DC: Education Policy Institute. Retrieved January 3, 2019, from https://www.epi.org/publication/the-teacher-pay-gap-is-wider-than-ever-teachers-pay-continues-to-fall-further-behind-pay-of-comparable-workers/

Allen, R., & Sims, S. (2017). *Improving Science Teacher Retention: Do National STEM Learning Network professional development courses keep science teachers in the classroom?* Wellcome and Education Datalab. Retrieved from https://wellcome.ac.uk/sites/default/files/science-teacher-retention.pdf

Allen, R., & Sims, S. (2018). *The teacher gap*. UK: Routledge.

Ambrosetti, A., & Dekkers, J. (2010). The interconnectedness of the roles of mentors and mentees in pre-service teacher education mentoring relationships. *Australian Journal of Teacher Education, 35*(6), 42–55.

Antink-Meyer, A., & Brown, R. A. (2017). Second-career science teachers' classroom conceptions of science and engineering practices examined through the lens of their professional histories. *International Journal of Science Education, 39*(11), 1511–1528.

Archer, L., Dawson, E., DeWitt, J., Seakins, A., & Wong, B. (2015). 'Science capital': A conceptual, methodological, and empirical argument for extending bourdieusian notions of capital beyond the arts. *Journal of Research in Science Teaching, 52*(7), 922–948.

Archer, L., DeWitt, J., Osborne, J., Dillon, J., Willis, B., & Wong, B. (2013). 'Not girly, not sexy, not glamorous': Primary school girls' and parents' constructions of science aspirations. *Pedagogy, Culture & Society, 21*(1), 171–194.

Atteberry, A., Loeb, S., & Wyckoff, J. (2016). Teacher churning: Reassignment rates and implications for student achievement. *Educational Evaluation and Policy Analysis, 20*(10), 3–30.

Beijaard, D., Verloop, N., & Vermunt, J. D. (2000). Teachers' perceptions of professional identity: An exploratory study from a personal knowledge perspective. *Teaching and Teacher Education, 16*(7), 749–764.

Bell, M., & Cordingley, P. (2014). *Characteristics of high performing schools*. UK: Centre for use of research and Evidence in Education. Retrieved June 27, 2020, from http://www.curee.co.uk/files/publication/%5Bsite-timestamp%5D/Teach%20First%20Research%20Report.pdf

Bennett, J., Dunlop, L., Knox, K. J., Reiss, M. J., & Torrance Jenkins, R. (2018). Practical Independent Research Projects in science: A synthesis and evaluation of the evidence of impact on high school students. *International Journal of Science Education, 40*(14), 1755–1773.

Binkhorst, F., Handelzalts, A., Poortman, C. L., & Van Joolingen, W. R. (2015). Understanding teacher design teams—A mixed methods approach to developing a descriptive framework. *Teaching and Teacher Education, 51*, 213–224.

Bourdieu, P. (1977). *Outline of a theory of practice* (Vol. 16). Cambridge, UK: Cambridge University Press.

Bueno, C., & Sass, T. (2016). The effects of differential pay on teacher recruitment, retention and quality. In 2016 Fall Conference: *The Role of Research in Making Government More Effective*. Retrieved November 3, 2019, from https://edre.uark.edu/lecture-series/effectsofdifferentialpaysass.pdf

Chung-Parsons, R., & Bailey, J. M. (2019). The hierarchical (not fluid) nature of preservice secondary science teachers' perceptions of their science teacher identity. *Teaching and Teacher Education, 78*, 39–48.

Cordingley, P., Greany, T., Crisp, B., Seleznyov, S., Bradbury, M., & Perry, T. (2018). *Developing great subject teaching*. London: Wellcome. Retrieved September 2, 2019, from https://wellcome.ac.uk/sites/default/files/developing-great-subject-teaching.pdf

Danielowich, R. M. (2012). Looking through different lenses: How preservice science teachers use practice-oriented reflections to negotiate more reform-minded identities. *Journal of Science Teacher Education, 23*(4), 323–346.

Darling-Hammond, L., Hyler, M. E., & Gardner, M. (2017, June). *Effective teacher professional development*. Learning Policy Institute. Retrieved October 26, 2017, from http://learningpolicyinstitute.org/sites/default/files/product-files/Effective_Teacher_Professional_Development_REPORT.pdf

Darling-Hammond, L., & Richardson, N. (2009). Teacher learning: What matters? *Educational Leadership, 66*(5), 46–53.

Davies, P., Connolly, M., Nelson, J., Hulme, M., Kirkman, J., & Greenway, C. (2016). 'Letting the right one in': Provider contexts for recruitment to initial teacher education in the United Kingdom. *Teaching and Teacher Education, 60*, 291–302.

Department for Education (DfE). (2018). *School workforce in England: November 2018*. Statistical First Release. Retrieved October 10, 2019, from https://www.gov.uk/government/statistics/school-workforce-in-england-november-2018

Department for Education (DfE). (2019a). *Teacher recruitment and retention strategy*. Retrieved June 20, 2019, from https://www.gov.uk/government/publications/teacher-recruitment-and-retention-strategy

Department for Education (DfE). (2019b). *Initial Teacher Training (ITE) census for 2019 to 2020. Statistical first release*. Retrieved June 27, 2020, from https://assets.publishing.service.gov.uk/government/uploads/system/uploads/attachment_data/file/848851/ITE_Census_201920_Main_Text_final.pdf

DeWITE, J., & Archer, L. (2015). Who aspires to a science career? A comparison of survey responses from primary and secondary school students. *International Journal of Science Education, 37*(13), 2170–2192.

El Nagdi, M., Leammukda, F., & Roehrig, G. (2018). Developing identities of STEM teachers at emerging STEM schools. *International Journal of STEM Education, 5*(36). https://doi.org/10.1186/s40594-018-0136-1

Eldridge, D., Lock, R. & Vokins, M. (1995). *Biology: Projects and investigations*. Nuffield New Science. Longman.

European Commission/EACEA/Eurydice. (2018). *Teaching careers in Europe: Access, progression and support*. Eurydice Report. Luxembourg: Publications Office of the European Union.

Forni, A. P. (2007). *Shortage of qualified science teachers in Norway. A system dynamics approach*. Unpublished Masters' Thesis, University of Bergen. Retrieved July 7, 2019, from http://bora.uib.no/handle/1956/3274

Foster, D. (2018). *Teacher recruitment and retention in England*. House of Commons briefing paper number 7222. Retrieved April 4, 2019, from https://researchbriefings.parliament.uk/ResearchBriefing/Summary/CBP-7222

Foster, D. (2019). *Initial teacher training in England*. House of Commons briefing paper number 6710. Retrieved November 3, 2019, from https://researchbriefings.files.parliament.uk/documents/SN06710/SN06710.pdf

Fredriksson, P., Öckert, B., & Oosterbeek, H. (2013). Long-term effects of class size. *The Quarterly Journal of Economics, 128*(1), 249–285.

Friedrichsen, P., Lannin, J., Abell, S., Arbaugh, F., & Volkmann, M. (2008). Examining incoming identities in an alternative certification program in

mathematics and science. *Eurasia Journal of Mathematics, Science & Technology Education, 4*(2), 175–188.
Furtak, E. M., Seidel, T., Iverson, H., & Briggs, D. C. (2012). Experimental and quasi-experimental studies of inquiry-based science teaching: A meta-analysis. *Review of Educational Research, 82*(3), 300–329.
García, E., & Weiss, E. (2019). *The teacher shortage is real, large and growing, and worse than we first thought.* Economic Policy Institute Report. Washington, DC: Economic Policy Institute. Retrieved from https://www.epi.org/163651
Gibson, H. L., & Chase, C. (2002). Longitudinal impact of an inquiry-based science program on middle school students' attitudes towards science. *Science Education, 86*, 693–705.
Godec, S., King, H., & Archer, L. (2017). *The science capital teaching approach: Engaging students with science, promoting social justice.* London: University College London.
Grier, J. M., & Johnston, C. C. (2009). An inquiry into the development of teacher identities in STEM career changers. *Journal of Science Teacher Education, 20*(1), 57–75.
Gusky, T. R. (2002). Professional development and teacher change. *Teachers and Teaching: Theory and Practice, 8*(3/4), 381–391.
Hattie, J. A. C. (2009). *Visible learning: A synthesis of 800+ meta-analyses on achievement.* London: Routledge.
Henry, G. T., Fortner, C. K., & Bastian, K. C. (2012). The effects of experience and attrition for Novice high school science and mathematics teachers. *Science, 335*(6072), 1118–1121.
Heredia, S. C., & Yu, J. H. (2017). A matter of choice: Opportunities for informal science institutions to support science teacher induction. *Journal of Science Teacher Education, 28*(6), 549–565.
Human Developing Index (HDI). (2006). *United Nations Human Developing Index.* Retrieved November 3, 2019, from http://hdr.undp.org/hdr2006/statistics
Irving-Bell, D. M. (2018). *The formation of science, technology, engineering and mathematics teacher identities: Pre-service teachers' perceptions.* Unpublished doctoral dissertation. Retrieved from https://eprints.lancs.ac.uk/id/eprint/127684/1/2018DawneMarilynIrving_BellPhD.pdf
Jackson, D., & Temperley, J. (2007). From professional learning community to networked learning community. In L. Stoll & K. Seashore Louis (Eds.), *Professional learning communities: Divergence, depth and dilemmas* (pp. 45–62). England: McGraw Hill Open University Press.

Johnson, S. M., Kraft, M. A., & Papay, J. P. (2012). How context matters in high-need schools: The effects of teachers' working conditions on their professional satisfaction and their students' achievement. *Teachers College Record, 114*(10), 1–39.

Joyce, B., & Showers, B. (1988). *Student achievement through staff development*. New York: Longman.

Kennedy, A. (2014). Understanding continuing professional development: The need for theory to impact on policy and practice. *Professional Development in Education, 40*(5), 688–697.

Kennedy, M. M. (2016). How does professional development improve teaching? *Review of Educational Research, 86*(4), 945–980.

Lofthouse, R. M. (2018). Re-imagining mentoring as a dynamic hub in the transformation of initial teacher education: The role of mentors and teacher educators. *International Journal of Mentoring and Coaching in Education, 7*(3), 248–260.

Lofthouse, R. M. (2019). Coaching in education: A professional development process in formation. *Professional Development in Education, 45*(1), 33–45.

Lofthouse, R. M., & Leat, D. (2013). An activity theory perspective on peer coaching. *International Journal of Mentoring and Coaching in Education, 2*(1), 8–20.

Luehmann, A. (2016). Practice-linked identity development in science teacher education: GET REAL! Science as a figured world. In L. Avraamidou (Ed.), *Studying science teacher identity: Theoretical, methodological and empirical explorations* (pp. 15–48). Rotterdam: Sense.

Luehmann, A. L. (2007). Identity development as a lens to science teacher preparation. *Science Education, 91*(5), 822–839.

Machin, S., & Murphy, S. (2011). *Improving the impact of teachers on pupil attainment in the UK: Interim findings*. London: Sutton Trust. Retrieved June 6, 2017, from https://www.suttontrust.com/wp-content/uploads/2011/09/2teachers-impact-report-final.pdf

Manning, A. (2017). *Urban science teachers exploring how their views and experiences can influence decisions to remain in post or not*. Unpublished doctoral dissertation. Retrieved from https://core.ac.uk/download/pdf/141244962.pdf

McIntyre, J., & Hobson, A. J. (2016). Supporting beginner teacher identity development: External mentors and the third space. *Research Papers in Education, 31*(2), 133–158.

Mehalik, M. M., Doppelt, Y., & Schunn, C. D. (2008). Middle-school science through design-based learning versus scripted inquiry: Better overall science concept learning and equity gap reduction. *Journal of Engineering Education, 97*(1), 71–85.
Mehli, H., & Bungum, B. (2013). A space for learning: How teachers benefit from participating in a professional community of space technology. *Research in Science & Technological Education, 31*(1), 31–48.
Migration Advisory CommITEee (MAC). (2016). Partial review of the shortage occupation list: Review of teachers. Migration Advisory CommITEee. Retrieved November 3, 2019, from https://www.gov.uk/government/consultations/migration-advisory-commitee-mac-review-on-teacher-shortages
Minner, D. D., Levy, A. J., & Century, J. (2010). Inquiry-based science instruction—What it is and does it matter? Results from a research synthesis years 1984 to 2002. *Journal of Research in Science Teaching, 47*(4), 474–496.
Mizell, H. (2010). *Why professional development matters.* Learning forward [online]. Retrieved June 10, 2018, from https://learning forward.org/docs/pdf/why_pd_matters_web.pdf
Mocetti, S. (2012). Educational choices and the selection process: Before and after compulsory schooling. *Education Economics, 20*(2), 189–209.
Molander, B. O., & Hamza, K. (2018). Transformation of professional identities from scientist to teacher in a short-track science teacher education program. *Journal of Science Teacher Education, 29*(6), 504–526.
Mujtaba, T., & Reiss, M. J. (2013). Inequality in experiences of physics education: Secondary school girls' and boys' perceptions of their physics education and intentions to continue with physics after the age of 16. *International Journal of Science Education, 35*(11), 1824–1845.
NAO. (2016). *Training new teachers.* UK: National Audit Office Report. Retrieved November 1, 2019, from https://www.nao.org.uk/report/training-new-teachers/
NAO. (2017). *Retaining and developing the teacher workforce.* UK: National Audit Office Report. Retrieved September 1, 2019, from https://www.nao.org.uk/report/supporting-and-improving-the-teaching-workforce/
Nieswandt, M., Barrett, S. E., & McEneaney, E. H. (2013). Predictors of science subject discipline identities: A statistical analysis. *Canadian Journal of Science, Mathematics and Technology Education, 13*(1), 90–110.
O'Doherty, T., & Hartford, J. (2018). Teacher recruitment: Reflections from Ireland on the current crisis in teacher supply. *European Journal of Teacher Education, 41*(5), 654–669.

OECD. (2015). *Universal basic skills: What countries stand to gain*. Paris: OECD Publishing.
OECD. (2018). Effective teacher policies: Insights from PISA. https://doi.org/10.1787/9789264301603-en. In P. Sahlberg, P. Munn, & J. Furlong (Eds.), 2012. *Report of the International Review Panel on the Structure of Initial Teacher Education Provision in Ireland.* Dublin: Department of Education and Skills.
Ornstein, A. (2006). The frequency of hands-on experimentation and student attitudes toward science: A statistically significant relation. *Journal of Science Education and Technology, 15*(3–4), 285–297.
Ortega, C. M. V., Correa Molina, E., & Fuentealba Jara, A. R. (2014). La práctica del profesor de Ciencias: Significados personales y experiencias de profesores en formación. *Perspectiva Educacional, 54*(1), 17–34.
Papay, J. P., & Kraft, M. A. (2015). Productivity returns to experience in the teacher labor market: Methodological challenges and new evidence on long-term career improvement. *Journal of Public Economics, 130*, 105–119.
Proweller, A., & Mitchener, C. P. (2004). Building teacher identity with urban youth: Voices of beginning middle school science teachers in an alternative certification program. *Journal of Research in Science Teaching, 41*(10), 1044–1062.
Reiss, M. J. (1992). School ecology practicals. In: D. Harding (Complier) Ecology Projects. *Ideas and practicals from the Journal of biological education* (pp. 1–2). Institute of Biology.
Richmond, G. (2016). Making sense of the interplay between identity, agency and context in the development of beginning science teachers in high-poverty schools. In L. Avraamidou (Ed.), *Studying science teacher identity: Theoretical, methodological and empirical explorations* (pp. 219–238). Rotterdam: Sense.
Rivera Maulucci, M. S. (2013). Emotions and positional identity in becoming a social justice science teacher: Nicole's story. *Journal of Research in Science Teaching, 50*(4), 453–478.
Rivera Maulucci, M. S., & Fann, K. T. (2016). Teaching for social justice in science education: Helping a new teacher develop a social justice identity. In L. Avraamidou (Ed.), *Studying science teacher identity: Theoretical, methodological and empirical explorations* (pp. 111–128). Rotterdam: Sense.
Rivkin, S. G., Hanushek, E. A., & Kain, J. F. (2005). Teachers, schools and academic achievement. *Econometrica, 73*(2), 417–458.

Rockoff, J. E. (2004). The impact of individual teachers on student achievement: Evidence from panel data. *American Economic Review Papers and Proceedings, 94*(2), 247–252.

Ronfeldt, M., Loeb, S., & Wyckoff, J. (2012). How teacher turnover harms student achievement. *American Educational Research Journal, 50*(1), 4–36.

Rushton, E. A. C., Charters, L., & Reiss, M. J. (2019). The experiences of active participation in academic conferences for high school science students. *Research in Science and Technological Education*. https://doi.org/10.1080/02635143.2019.1657395

Rushton, E. A. C., & Parker, B. (2019). Empowering young people to develop STEM careers through active participation in genuine scientific research. In S. E. Hiller & A. Kitsantas (Eds.), *Citizen Science Programs: Guidelines for informal science educators in enhancing youth science motivation and achievement* (pp. 97–127). Hauppauge, NY: Nova Science Publishers, Inc..

Rushton, E. A. C., & Reiss, M. J. (2019). From science teacher to 'teacher scientist': Exploring the experiences of research-active science teachers in the UK. *International Journal of Science Education, 41*(11), 1541–1561.

Rushton, E. A. C., & Reiss, M. J. (2020). Middle and high school science teacher identity considered through the lens of the social identity approach: A systematic review of the literature. *Studies in Science Education*. https://doi.org/10.1080/03057267.2020.1799621

Schaap, H., Louws, M., Meirink, J., Oolbekkink Marchand, H., Van Der Want, A., Zuiker, I., … Meijer, P. (2019). Tensions experienced by teachers when participating in a professional learning community. *Professional Development in Education, 45*(5), 814–831.

Sheldrake, R., Mujtaba, T., & Reiss, M. J. (2017a). Students' changing attitudes and aspirations towards physics during secondary school. *Research in Science Education, 49*, 1809–1834.

Sheldrake, R., Mujtaba, T., & Reiss, M. J. (2017b). Science teaching and students' attitudes and aspirations: The importance of conveying the applications and relevance of science. *International Journal of Educational Research, 85*, 167–183.

Sibieta, S. (2018). *The teacher labour market in England. Shortages, subject expertise and incentives*. London: Education Policy Institute. Retrieved from https://dera.ioe.ac.uk/32106/1/EPI-Teacher-Labour-Market_2018-1.pdf

Sims, S. (2017). *What happens when you pay shortage-subject teachers more money? Simulating the effect of early-career salary supplements on teacher supply in England*. London: Gatsby Foundation.

Sims, S. (2019). *Increasing the quantity and quality of science teachers in schools: Eight evidence-based principles*. London: Gatsby Foundation.

Skolverket. (2017). *Pedagogisk personal i skola och vuxenutbildning läsåret 2016/17* [Pedagogic staff in school and adult education in the school year 2016/17]. PM 2017:370. Stockholm: The Swedish National Agency for Education.

Smithers, A., & Robinson, P. (2000). *Coping with teacher shortages*. Centre for Education and Employment Research. Retrieved June 27, 2020, from https://pdfs.semanticscholar.org/76ad/c389d236d0ddedf19fd84ae9e99f6654d594.pdf

Snyder, C., Oliveira, A. W., & Paska, L. M. (2013). STEM career changers' transformation into science teachers. *Journal of Science Teacher Education, 24*(4), 617–644.

Stock Jones, R., Annable, T., Billingham, Z., & MacDonald, C. (2016). *Quantifying CREST. CREST Silver Award Evaluation*. A Probono Economics Research Report for the British Science Association. Retrieved from https://www.britishscienceassociation.org/Handlers/Download.ashx?IDMF=9aaf06e4-f24a-435c-9991-cbf6edd1293c

Stoet, G., & Geary, D. C. (2018). The gender-equality paradox in science, technology, engineering, and mathematics education. *Psychological Science, 29*(4), 581–593.

Sutcher, L., Darling-Hammond, L., & Carver-Thomas, D. (2016). *A coming crisis in teaching? Teacher supply, demand and shortages in the U.S.* Learning Policy Institute Report. Retrieved from https://learningpolicyinstitute.org/sites/default/files/product-files/A_Coming_Crisis_in_Teaching_REPORT.pdf

Teach First. (2019). *Our impact*. Teach First website. Retrieved November 3, 2019, from https://www.teachfirst.org.uk/our-impact

The Brilliant Club. (2020). *About the brilliant club*. Brilliant Club website. Retrieved March 31, 2020, from https://thebrilliantclub.org/about-the-brilliant-club/what-is-tbc/

Tschannen-Moran, B., & Tschannen-Moran, M. (2010). *Evocative coaching* (1st ed.). San Francisco: Juliessey-Bass.

Ufnar, J. A., Bolger, M., & Shepherd, V. L. (2017). A retrospective study of A Scientist in the Classroom Partnership program. *Journal of Higher Education Outreach and Engagement, 21*(3), 69–96.

Ufnar, J. A., Kuner, S., & Shepherd, V. L. (2012). Moving beyond GK-12. *CBE Life Science Education, 11*(3), 239–247.

Ufnar, J. A., & Shepherd, V. L. (2019). The Scientist in the Classroom Partnership program: An innovative teacher professional development model. *Professional Development in Education, 45*(4), 642–658.

Varelas, M., House, R., & Wenzel, S. (2005). Beginning teachers immersed into science: Scientist and science teacher identities. *Science Education, 89*(3), 492–516.

Veenman, S., & Denessen, E. (2001). The coaching of teachers: Results of five training studies. *Educational Research and Evaluation, 7*(4), 385–417.

Watters, J. J., & Diezmann, C. M. (2015). Challenges confronting career-changing beginning teachers: A qualitative study of professional scientists becoming science teachers. *Journal of Science Teacher Education, 26*(2), 163–192.

Weldon, P. (2018). Early career teacher attrition in Australia: Evidence, definition, classification and measurement. *Australian Journal of Education, 6*(1), 61–78.

West, R. J. (2015). *Teacher identity and self-efficacy development in an alternative licensure program for middle and high school math and science teachers.* Unpublished doctoral dissertation. Retrieved from http://rave.ohiolink.edu/etdc/view?acc_num=ohiou1427381851

Wilson, E., & Deaney, R. (2010). Changing career and changing identity: How do teacher career changers exercise agency in identity construction? *Social Psychology of Education, 13*(2), 169–183.

Wolstenholme, C., Coldwell, M., & Stevens, A. (2012). *The impact of Science Learning Centre continuing professional development on teachers' retention and careers: Final report.* Sheffield: CEIR.

Woolhouse, C., & Cochrane, M. (2015). Educational policy or practice? Traversing the conceptual divide between subject knowledge, pedagogy and teacher identity in England. *European Journal of Teacher Education, 38*(1), 87–101.

Worth, J., De Lazzari, G., & Hillary, J. (2017). *Teacher retention and turnover research: Interim report.* Slough: NFER.

Yoon, K. S., Duncan, T., Lee, S. W. Y., Scarloss, B., & Shapley, K. (2007) *Reviewing the evidence on how teacher professional development affects student achievement* (Issues & Answers Report, REL 2007-No. 033). Washington, DC: U.S. Department of Education, Institute of Education Sciences, National Centre for Education Evaluation and Regional Assistance, Regional Educational Laboratory Southwest. Retrieved from https://ies.ed.gov/ncee/edlabs/regions/southwest/pdf/REL_2007033.pdf

2

Science Teacher Identity

2.1 Exploring the Formation of High School Science Teacher Identity and the Social Identity Approach

Understanding the role of identity in shaping the experiences of high school science teachers is important and timely, given the challenging global context of teacher recruitment and retention (particularly chemistry and physics specialists) that has been outlined in Chap. 1. As Rushton and Reiss (2020) have described, researchers have explored the place of identity in broad considerations of teacher identity (Beauchamp & Thomas, 2009; Beijaard, Meijer, & Verloop, 2004; Beijaard, Verloop, & Vermunt, 2000; Gee, 2000; Izadinia, 2013) as well as in research with a focus on science teacher identity (Avraamidou, 2014, 2016; Glass, 2019). Avraamidou (2014) drew together ways in which the general concept of identity has been used as a lens to explore a range of aspects of science education including science learning, teacher preparation, teacher identity (reform-minded teacher identity; subject matter knowledge; competence, performance and recognition; life histories and context) and teacher identity development (participation in field-based courses;

E. A. C. Rushton, *Science Education and Teacher Professional Development*, Palgrave Studies in Alternative Education, https://doi.org/10.1007/978-3-030-64107-8_2

informal science experiences; use of technology applications; use of curriculum materials; personal histories and biographies). Although Avraamidou (2014) demonstrated that identity has been used extensively as a lens in education, the social identity approach has had limited application in education (Mavor, Platow, & Bizumic, 2017) and has been all but overlooked in considerations of the identity of science teachers (Rushton & Reiss, 2019). The social identity approach (Haslam, Reicher, & Platow, 2011) brings together theories of shared or group identity (Tajfel & Turner, 1979) and self-categorisation (Turner, Hogg, Oakes, Reicher, & Wetherell, 1987) to explain the role that social context has in determining an individual's sense of self and identity. This chapter has three aims. Firstly, to bring together the Social Identity Approach (SIA) with a focus on science teacher identity formation by describing the theoretical framework of the SIA and the application of these concepts to the field of education. Secondly, to provide a broad overview of identity in literature focused on science teachers and, thirdly, to consider in more detail an aspect of identity formation, namely school environments, that have yet to be considered in Chap. 1. The second and third aims of this chapter are met by drawing on the findings of Rushton and Reiss' (2020) systematic review of studies that report on findings from research into middle and/or high school science teacher identity, conducted within the last 20 years. Specifically, Rushton and Reiss (2020) analysed the findings of 79 empirical and/or theoretical publications to consider to what extent and which was the SIA might provide a useful lens through which to consider middle and/or high school science teacher identity and how this informs identity formation. Before considering aspects of Rushton and Reiss (2020), the theoretical framework of SIA is described with particular reference to education.

2.2 Theoretical Framework

2.2.1 Social Identity Approaches and Education

Social identity approaches combine the theories of *social identity* (Tajfel & Turner, 1979) and *self-categorisation* (Turner et al., 1987) which together are referred to as the *social identity approach* (Haslam et al., 2011). These approaches suggest that a person's sense of self is largely determined by their social context and the groups to which they belong and identify with, and that people seek to develop and maintain a positive view of themselves by comparing themselves and their group memberships in a more positive light than their alternative 'outgroups' (Ellemers & Haslam, 2012; Tajfel & Turner, 1979). Social identity theory suggests how groups form (Turner, 1982) whereas self-categorisation theory suggests when groups form, and recognises that this is part of a context-sensitive, self-categorisation process (Oakes, Haslam, & Turner, 1994; Turner, 1982).

Education is a collaborative process that involves groups of people and yet the SIA has only recently been considered in relation to education research (Mavor et al., 2017). In a review of current research, Haslam (2017) identifies that five 'I's that have significance for social identity and education: *Identification, Ideation*, *Interaction*, *Influence* and *Ideology*. *Identification* is based upon the idea that group membership shapes an individual's behaviour to the extent that their social identity derived from this group membership is incorporated into their sense of self. Research considering identification and teachers has shown that levels of identification are good predictors of engagement (Christ, van Dick, Wagner, & Stellmacher, 2003), job satisfaction (van Dick & Wagner, 2001) and self-reported physical (van Dick & Wagner, 2002) and psychological (van Dick & Wagner, 2001) health and well-being. The concept of *Ideation*, highlights that what people identify with, is as important as mutual identification. For example, what students are told about the groups they identify with shapes their behaviour in both school (Boucher & Murphy, 2017; Mujtaba & Reiss, 2013) and university (Cruwys, Gaffney, & Skipper, 2017) settings. *Interaction* is what develops and galvanises social identities (Haslam, 2017) and this interaction has the capacity to shape the extent to which individuals feel part of the group and therefore can increase or limit their academic and intellectual performance (Reynolds, Subasic, Bromhead, & Lee, 2017). *Influence*, Haslam (2017) suggests, is what makes identification, ideation and interaction possible; it is the extent to which leaders can shape the attitudes, intentions and behaviour of followers. Levels of influence are determined by how much followers socially identify with their leader. *Ideology* pervades education, and multiple aspects of the educational landscape (e.g. class, political views, gender, 'race', faith) provide teachers with the context for identification, ideation and interaction (Haslam, 2017).

Thus far, studies that consider ideation and interaction in education have focused on student identities, including studies that explore female students' experiences of STEM subjects at school and in higher education (Boucher & Murphy, 2017), bulling (Jones, Livingstone, & Manstead, 2017) and socio-economic factors relating to educational experience (Jetten, Iyer, & Zhang, 2017). Rushton and Reiss (2019) identified that, apart from the work of Mavor et al. (2017), social identity perspectives have yet to feature significantly in research focused upon teachers (or technicians) and, that the systematic review of the literature (Rushton & Reiss, 2020) and this current study, seek to further contribute to this gap in current understanding. Drawing on the research of Mavor et al. (2017) and Rushton and Reiss (2020), I argue that the SIA is appropriate for science education because promoting social connectedness and integration, so that those involved in teaching have a shared sense of 'us', will further support education that is meaningful, purposeful and effective for all participants.

A detailed methodology is described in Rushton and Reiss (2020). In summary, the review of Rushton and Reiss (2020) took the form of a systematic review, with four stages which identified five groups of studies including:

1. studies that broadly consider identity development in teachers, using a variety of theories, frameworks and models;
2. studies that focus on a specific aspect of a teacher's own context or experience, such as cultural contexts and career changers;

3. studies that focus on an aspect of the school and/or teaching context, including teaching inquiry and new curricula;
4. studies that consider the role of the training programme, for example, those that are based in alternative institutions or situated in reform-minded or social justice frameworks that are explicit; and
5. studies that consider the role of professional development opportunities, including leadership, STEM research and blogging.

Findings from Rushton and Reiss (2020) linked to teacher professional development, including alternative training programmes were incorporated into Chap. 1 whilst, in Chap. 2 research that broadly considers the role of identity formation in science teachers and, more specifically, the role of school environments are now considered.

2.2.2 Conceptual Frameworks and Models Utilised to Underpin Understandings of 'Identity'

Rushton and Reiss (2020) highlighted the complexity of identity as a concept; however, common to all studies was a recognition that identity was a dynamic, multidimensional and fluid construct that was universally recognised to change over time. Rushton and Reiss (2020) identified two conceptual models of science identity found in the literature: The Dynamic Systems Model of Role Identity (DSMRI) (Garner, Hathcock, & Kaplan, 2016; Garner & Kaplan, 2019) and The Science Teacher Identity (STI) (Chi, 2009). Both models referenced Gee's (2000) understanding of the four types of identity that are socially and culturally constructed by nature, institutions, discourse and affinity. The DSMRI has teacher learning and teacher identity development as integral processes with four interrelated components, guided by the cultural context, contributing to the role of a teacher: (1) ontological and epistemological beliefs; (2) purpose and goals; (3) self-perceptions and self-definitions and (4) perceived action and future possibilities (Garner et al., 2016; Garner & Kaplan, 2019). Garner and Kaplan (2019) suggested that the DSMRI provides a teacher with a framework through which to interpret their experiences and enact their role, developing their identity in a dynamic and iterative process. Applying the DSMRI to a case study of a veteran high school science teacher, Garner et al. (2016) argued that three role identities were enacted during the professional development programme: teacher, learner and scientist or content expert (Garner et al., 2016). Understandings drawn from the application of the DSMRI suggest that teachers' learning within professional development contexts is dynamic, complex and is shaped by the design and delivery of the professional development programme (Garner et al., 2016; Garner & Kaplan, 2019). Chi (2009) developed a model of Science Teacher Identity (STI) with nine dimensions: personal learning experience; knowledge and skills; community practice; science teaching practice; degree of success; social respect; belief and value in science teaching; intrinsic satisfaction and representation. Chi (2009) created a 48-item questionnaire to provide a score of STI and argued that this instrument could be used to measure changes in science teacher identity, for example, before and after a teacher's participation in professional development programme. As one would expect, STI scores were positively associated with increased teaching experience, knowledge and skills, and participation in professional development opportunities (Chi, 2009). The nine dimensions of the STI suggested by Chi are closely related to the four concepts of the DSMRI (Garner et al., 2016; Garner & Kaplan, 2019). For example, the dimensions of personal learning experience, belief, and values and representation from the STI (Chi, 2009) could incorporate the same material found in the ontological and epistemological beliefs component of the DSMRI (Garner et al., 2016; Garner & Kaplan, 2019). Likewise, the dimensions of science teaching practice and degree of success (Chi, 2009) could be understood as part of a teachers' perceived actions and future possibilities (Garner et al., 2016; Garner & Kaplan, 2019). However, a difference between these two models is that Chi (2009) explicitly describes social and community aspects to such dimensions (community practice, social respect), whereas this is only implied in the four concepts of the DSMRI (Garner et al., 2016; Garner & Kaplan, 2019). Rushton and Reiss (2020) contend that the social and community aspects of identity should be an explicit part of a middle and/or high school science teacher identity model, as the SIA recognises that the identity derived from group membership is incorporated into an individual's sense of self and shapes their behaviour.

Rushton and Reiss (2020) found that the work of Gee (2000) was explicitly incorporated in the conceptual frameworks of three further studies that were focused on science teachers categorised as 'exemplary' (Deneroff, 2016), 'preservice' (Hsu, Reis, & Monarrez, 2017) and 'experienced' (Richardson, 2019). In a three-year ethnographic study of the identity construction and teaching practices of an urban high school science teacher, who is described as being an exemplary teacher of inquiry, Deneroff (2016) drew on Gee (2000) to suggest that science teaching is not the summation of knowledge, values and beliefs but should be understood as being socially constructed and that this understanding should inform the design of professional development programmes for teachers that enable teachers to consciously problematise identities if they are to transform their teaching practice. Both research that is focused on preservice middle school science teachers (Hsu et al., 2017) and that which examines experienced high school science teachers (Richardson, 2019) drew from Gee (2000) the concept of 'affinity', which Hsu et al. (2017) articulated as 'shared practices with group members'. In an analysis of written teaching philosophies and learning autobiographies from 38 preservice middle school science teachers, written as part of their final year of a four-year university-based teacher education programme, Hsu et al. (2017) found that affinity identity was the most important factor in shaping science learning experiences and preservice science teachers' teaching philosophies. However, affinity identities were mediated in slightly different ways. When preservice teachers discussed their affinity identities as learners, they emphasised the importance of the social context of the classroom and the collective practices they had experienced. However, when the same preservice teachers discussed their affinity identities as future science teachers, they focused more closely on their own individual characteristics as teachers. In a study that considered the reasons why science teachers remain in the profession, Richardson (2019) also recognised the importance of affinity identity, and referenced the work of Gee (2000) to assert that teachers who are aware of their professional identities are more likely to be retained in the profession. In an examination of the experiences of 20 mid-and late-career high school science teachers, Richardson (2019) suggested that affinity identity is more important for mid-career teachers' identity development, with this cohort of teachers placing greater importance on developing connections with colleagues than late-career teachers, whose institutional identity held more significance. Richardson (2019) suggested that this is most likely due to late-career teachers having greater contact with authority figures within and beyond their schools and because they often held leadership roles themselves.

Other frameworks that have been utilised by researchers and are consistent with Gee's (2000) identity theory that recognises that an individual's identity is shaped by their social context include the tripartite division of identity by Day, Kington, Stobart, and Sammons (2006). Day et al. (2006) identified facets including (1) professional—what constitutes a good teacher, (2) situated—the school context and (3) personal—life outside of school, with each having competing sub-identities that are held in balance or imbalance. In her exploration of five urban high school science teachers' experiences over a four-year period, Manning (2017) argued that teachers' personal and situated identities were more influential in determining teacher retention or attrition than professional identities. Specifically, teachers' personal resilience, notions of choice over whether to leave their role and teacher perceptions of their ability and the school's ability to mediate professional concerns (e.g. workload). Manning (2017) noted that it was somewhat surprising that professional identity was the least significant of these three components of identity, particularly when it is often government policy, rather than school policy or individual choice that determines issues such as workload and accountability measures. The importance of pragmatism (i.e. teachers' choices to stay or leave their roles often being determined by factors such as needing employment to meet financial commitments, rather than job satisfaction) is recognised by Manning (2017) in high school science teachers when they are negotiating challenging contexts. This understanding of pragmatic identities is arguably more consistent with Dominguez et al.'s (2015) view of interacting objective (e.g. constraints of a teacher education programme) and subjective (e.g. individuals' previous experiences) identities rather than Day et al.'s (2006) delineated, yet overlapping, personal, situated and professional identities. Nichols, Schutz, Rodgers and

Bilica (2017) explored identity development through a framework of emotions, and argued that emotional experiences that conflict with an individual's expectations of what it is to be a teacher provided opportunities for 'identity work', where teachers actively and explicitly explored what it is to be a teacher. This framework of emotions intersects an individual's personal and professional identity and is consistent with an understanding of professional identity development that pragmatically (Manning, 2017) interacts across both subjective and objective aspects of identity (Dominguez et al., 2015). This intersection is consistent with the conceptualisation of students' science identity (as opposed to science *teacher* identity) argued by Avraamidou (2019), who suggests that both emotions and recognition together are core, intertwined features of identity.

In their study of three preservice high school teachers' conceptions of their science teacher identities, Chung-Parsons and Bailey (2019) recognised that 'identities reflect the individual to include a sense of belonging to one or more larger social groups, such as the teaching profession' (p. 40). Through a consideration of the findings of Schachter and Rich (2011), Chung-Parsons and Bailey (2019) suggested that in order to facilitate students' development of science identity, teachers should 'identify with scientists and the scientific community, and have a science identity that manifests itself in their teaching identity' (p. 40). Chung-Parsons and Bailey (2019) found that teachers who undertook a year-long teacher accreditation course (often following a three-year undergraduate science-based degree) thus: (1) they viewed their science identity as distinct from their science teacher identity and that only their science identity was part of their core identity, (2) their science identity was only used in two contexts—teaching science content and analysing student work to facilitate learning—and (3) their teacher identity became dominant over their science identity only in the sociocultural context of science classrooms (Chung-Parsons & Bailey, 2019). In citing Lave (1996), Wenger (1998) and Burke and Stets (2009), Chung-Parsons and Bailey (2019) recognised the social context and reciprocal nature of identity development. However, what is omitted from their analysis of the dominance of the science identity over the science teacher identity is an understanding of the importance of group membership. Two of the three participants described themselves as 'science people' and did not identify that they wanted to become teachers until the end of the science degree. It is therefore unsurprising that they had not, over the course of a training year, yet developed science teacher identities through group membership that superseded well-established science identities.

Hong, Greene, and Lowery (2017) drew on the dialogical self-theory of Akkerman and Meijer (2011) to explore the development of high school science teacher identity in five individuals during a four-year study beginning at the preservice year until the end of their third year of teaching post-qualification. Hong et al. (2017) identified three themes from Akkerman and Meijer (2011): (1) multiplicity and unity—multiple identities held with a common thread; (2) social and individual—the role that the social context plays in the formation of individual identity and (3) discontinuity and continuity—change over time. In contrast to Day et al. (2006), where the focus of personal, situated and professional identities is broadly on the end-product of identity (e.g. mother, Head of Department, outstanding inquiry teacher), Hong et al.'s (2017) themes overtly focused on the contrasts within different aspects of identity. Hong et al. (2017) argued that the distinction between professional and personal identity is a false one, and that teacher identity development would be better understood without the separation of identity facets into these two broad groups and with a greater understanding of the importance of a supportive social environment. This understanding of the value of a supportive social environment is consistent with other research studies considered in this group (e.g. Manning, 2017; Richardson, 2019), although the importance of the social environment is not always given the same weight. In the following section, the role of the school environment in identity formation is considered.

2.2.3 Science Teacher Identity Formation and the School Environment

Studies that explore the identity development of science teachers with a focus on a specific aspect of the school environment and/or teaching context include studies focused on preservice and qualified teachers. Reading across these studies, Rushton and Reiss (2020) identified two foci that are particularly pertinent to this current study: (1) the school as a field or context of identity development and (2) the development of teachers' identity as teachers of inquiry.

Drawing on the work of the French sociologist, Pierre Bourdieu, to explore the social space or 'field' of a school science department, Melville, Wallace, and Bartley (2007) examined the relationships between ten experienced science teachers, almost all of whom had been teaching together for at least five years in the same Australian high school. Melville et al. (2007) argued that this approach helps elucidate the ways in which individuals identify as science teachers, drawing on personal and professional aspects developed in the social, interactive context of a science department. Rushton and Reiss (2020) argued that this conceptualisation of a science department as a distinct space or 'field' is helpful as it delineates the physical and ideological context for shared group membership. Teachers in this department shared professional identities as science department colleagues as well as science teachers in a way that would be recognised within the SIA through the concept of *influence*—where identity is shaped by the behaviour of group members, especially those understood as group leaders, in this case perhaps the head of the school science department. As described in Chap. 1, Bourdieu's concept of 'field' is an integral part of understandings about students' science capital—the field is both the physical setting (e.g. classroom) and social relationships (e.g. teacher-student) which frame the ways in which young people are able to benefit or not from science learning. Understanding the 'field' in which teachers develop identities as 'teacher scientists' is an important part of what this current study seeks to achieve.

Webb (2012) explored the identity formation of teachers in a discrete aspect of the 'field' of a school environment, namely, the induction programme and classroom science teaching of four first-year high school science teachers. Webb (2012) considered how these experiences shaped normative science teacher identities and found that through induction programmes, beginning science teachers developed identities that were more focused on school and school district policies and procedures compared to other aspects of science teaching, for example, inquiry-based instruction. This finding is consistent with a wider body of research (Luft,

2009; Luft et al., 2011; Luft & Zhang, 2014) that has explored the experiences of science teachers in different induction programmes. Rushton and Reiss (2020) argued that Webb's (2012) understanding of normative science teacher identities can be understood through the SIA as part of *ideation*, where group norms that are both descriptive ('what we do') and injunctive ('what we should do') shape behaviour. When people identify themselves as part of a group, they will enact the values and behavioural norms ascribed to that shared identity. Webb (2012) and the wider research literature (e.g. Luft, 2009; Luft & Zhang, 2014) have highlighted the importance of shared group membership for science teachers' well-being, with the close proximity of science specialists and school cultures found to be an important part of promoting beginning science teachers' well-being. This is core to the SIA, where social identity is central to positive personal and professional identity.

2.2.4 Teachers of Inquiry

Rushton and Reiss (2020) found that studies that considered teachers' identity in the context of inquiry-based teaching approaches were focused on preservice middle and high school science teachers (Dreon, 2008; Eick & Reed, 2002; Melville, Bartley, & Fazio, 2013). Bryce, Wilmes, and Bellino (2016) argued that professional development opportunities that promote inquiry-oriented teaching predominantly considered teaching practices rather than identity development. Bryce et al. (2016) called on the argument of Deneroff (2016) who highlighted the importance of including opportunities for reflexive transformation within professional development programmes. Methods and tools to promote such transformation in teachers included coaching, professional dialogues and online communities (Bryce et al., 2016), and Rushton and Reiss (2020) suggested that the importance of communities (both offline and online) underlines the value of group membership. Eick and Reed (2002) included the theoretical framework of personal histories to understand the development of preservice science teachers' identities and how this influences their ability to implement structured inquiry during their student teacher placement. Those teachers who identified themselves as 'constructors of knowledge' were more able to enact structured inquiry successfully than were those who saw themselves as 'traditional learners' who, through their student placements, were in receipt of knowledge about how to become teachers rather than contributing to knowledge or furthering understanding. Dreon (2008) observed that preservice middle and high school science teachers who enacted inquiry pedagogy experience emotions, of which anxiety was the predominant one observed. Preservice teachers experienced frequently unpredictable feelings of anxiety, arising in part from the level of their subject matter confidence, when delivering inquiry lessons. Preservice teachers' professional identities were also shaped by their interactions with their students and colleagues, and their perceived agency within the context of their schools' curricula and governance. Taken together, Dreon (2008) and Eick and Reed (2002) suggest that self-efficacy in subject content and teachers' own education experiences are all part of what enable beginning teachers to develop identities as teachers of inquiry. Rushton and Reiss (2020) suggested that these experiences and beliefs can be understood as part of the SIA, where teachers' ideation and ideologies inform their identity alongside shared interaction (e.g. with students and colleagues).

Melville et al. (2013) argued for the importance of scaffolding (i.e. where the learner gradually takes greater responsibility for learning with the diminishing direction of an expert) inquiry as an approach to teaching and learning to ensure preservice teachers appreciated the value of inquiry as a classroom strategy. When read in the context of Dreon (2008), this scaffolded approach could provide preservice teachers with the necessary support to engage in demanding professional activities, whilst reducing their anxiety. Melville et al. (2013) suggested that teachers should be actively engaged in constructing dual professional identities, both as students of science and as preservice teachers of science so that they have a robust framework when working within two unfamiliar contexts. Consistent with similar research studies, Melville et al. (2013) highlighted the importance of teachers' personal experiences (Eick & Reed, 2002) and emotions (Dreon, 2008) in constituting their own professional identities. In a 2-year study of 14 early-career US-based high school science teachers, Bang and Luft (2016) highlighted the importance of the school community in shaping beginning teachers' identities, and specifically in the context of inquiry-based teaching. Rushton and Reiss (2020) argued that how teachers are socially positioned directly relates to their ability to integrate into school settings and to how they develop their teaching identities. For example, teachers whose identity is positioned as a 'learner-of-being-a-teacher' have far greater opportunities to explore multiple identities, whereas teachers who are aligned as people who deliver a prescribed curriculum are less able to explore different pedagogical, intellectual, cultural and political spaces that contribute to the formation of a range of identities. Consistent with Bryce et al. (2016), Bang and Luft (2016) showed how online subject-specific mentoring provided science teachers with additional online and offline communities in which to develop agency in their role as teachers of inquiry learning. Common to these studies is a recognition that what teachers identify with (*ideation*) is important and can be shaped by their personal histories as well as their interactions with students and colleagues, both virtually and in person. A gap in the literature is the consideration of the identity formation of teachers who are experienced in inquiry, as the studies considered here are with preservice teachers or those in their first year of teaching post-qualification. Rushton and Reiss (2020) considered the identity formation of both experienced and early-career teachers who supported their students in inquiry-based, authentic learning over an extended period. Rushton and Reiss (2019) found that teachers with varying amounts of classroom experience developed positive, multifaceted professional identities. However, more research is needed with experienced teachers working in the context of inquiry learning to understand more fully how this teaching and learning approach shapes identity construction and through this current study, the role of inquiry in science teacher identity formation will be further explored in Chap. 9.

In Chap. 1 and this chapter, different aspects that influence science teacher identity formation including professional development, subject identity and school environments have been explored. Having given an overview of these areas it is important to reflect upon the relative strengths and weaknesses and future areas of development in the science teacher identity literature, and share what the SIA can offer to further understanding in this area.

2.3 Reflecting upon Literature Which Considers Science Teacher Identity Formation

Avraamidou (2014) identified limitations of the science teacher identity literature considered in her review. Key areas of weakness included (1) the lack of coherent and well-defined conceptualisation of science teacher identity; (2) the disconnect between theoretical frameworks and reform recommendations and (3) the need to understand how science identities are enacted in the classroom (Avraamidou, 2014). Methodological weaknesses included a reliance on very small participant numbers, studies that are a year or less in duration and where the context is oblique, and a predominance of qualitative studies (Avraamidou, 2014). Avraamidou (2014) contended that large-scale, quantitative studies would draw on a more inclusive and diverse participant group. That is not to say that qualitative studies do not bring substantial insights to understandings of identity. Indeed, their rich, narrative accounts enable reflective and detailed analysis and both Avraamidou's (2014) review and the systematic review of Rushton and Reiss (2020) confirm the valuable contributions that qualitative, narrative and case study approaches have made to this field of research.

Rushton and Reiss (2020) argued that in addition to greater clarification regarding context, researchers should seek to share the experiences of science identity development in non-Western cultural contexts, where English is not the first language. This would bring much needed diversity to the published academic literature which is still predominantly focused on studies from the USA. Relatedly, Rushton and Reiss (2020) contend that the literature would benefit from research that pertains to identity development of ethnic minority science teachers in their preservice training and throughout their teaching careers. As Seiler (2011) highlights, active science identities from non-dominant groups are frequently formed outside of school settings which, unlike their experiences of classrooms, do not conflict with their sense of self. Research in this area could contribute to studies that consider science identity development in students, including participation in science school and university courses as well as studies that consider public perceptions of science and scientists.

Avraamidou (2014) suggested five directions for research in science teacher identity:

1. studying science teacher identity as a process;
2. connecting science teacher identity research and reform recommendations;
3. conducting large-scale, longitudinal and life-history studies;
4. examining teacher identity enactment in school classrooms and
5. understanding the role of contexts on identity development.

Rushton and Reiss (2020) suggested that some advances have since been made in two of these research directions, namely (1) conducting large-scale, longitudinal and life-history studies; and (2) understanding the role of contexts of identity development. However, life-history and autobiographical studies (for example, Richardson, 2019; Roth, 2016), are rare. Rushton and Reiss (2020) also found that studies focused on experienced teachers (those with over ten years' teaching experience) were less frequent than those focused on preservice and beginning teachers. As is described in Chap. 3, of the 53 key informants who shared their experience through this research, 29 individuals had 12 or more years' teaching experience and the average was 13.9 years, which provides an important opportunity to explore the insights of teachers who can be described as 'experienced'.

The fifth future direction for science teacher identity research highlighted by Avraamidou (2014) was to consider the roles of different contexts on identity development. The study of context has been an area of growth in the literature, with studies exploring informal science institutions (Adams & Gupta, 2017); STEM workplaces (Antink-Meyer & Brown, 2017); online mentoring (Bang & Luft, 2016); blogging (Hanuscin, Cheng, Rebello, Sinha, & Muslu, 2014; Luehmann, 2008); STEM schools (El Nagdi, Leammukda, & Roehrig, 2018); research sites, for example, laboratories (Varelas, House, & Wenzel, 2005) and space technology research sites (Mehli & Bungum, 2013), and engagement in authentic science research (Rushton & Reiss, 2019). This current study draws on the context engagement in authentic science research to better understand the formation of science teacher identity.

Rushton and Reiss (2020) argued that the SIA provides a lens through which to consider a diverse literature. Drawing on Rushton and Reiss (2020), the final part of Chap. 2, focuses on the ways in which SIA gives a greater understanding of the importance of groups and group membership to teacher identity formation. As it is the focus of this current study, the research drawn on considers high school science teachers. However, I suggest that the arguments about group membership are also relevant for teachers of other disciplines and are therefore more widely applicable.

2.3.1 Group Membership and Social Identity: Understanding the Formation of High School Science Teacher Identity Through the Lens of the SIA

The SIA approach brings to the fore the importance of group life for an individual's health and well-being (Haslam, 2017). Rushton and Reiss (2020) argued that understanding the role and place of group membership and social identity in middle and high school science teacher identity development is core to providing those individuals with robust and purposeful professional identities. Rushton and Reiss (2020) demonstrated that the social context and practice of teaching has been acknowledged in the science teacher identity literature, but this coverage is patchy and at times contradictory. For example, one model of identity development explicitly recognises the community and social aspects to identity (Chi, 2009) whilst in another model the social component is only implied (e.g. Garner et al., 2016; Garner & Kaplan, 2019). Both Chi (2009) and Garner et al. (2016) drew on the work of Gee (2000) who, in identifying 'four ways to view identity' (p. 100), provides researchers with a concise conceptualisation of identity as 'a kind of person' (p. 99). In categorising four types of identity, Gee (2000) acknowledged that identity is socially constructed. Indeed, in his description of 'affinity identity', Gee (2000) highlighted that the source of power of this type of identity is experiences shared between 'affinity groups' (p. 100). Furthermore, when seeking to describe the concept of identity, Gee (2000) included the phrase 'being recognised as a kind of person' (p. 99). This is at the heart of the SIA: who a person is, their identity, is defined by the groups to which they are recognised by themselves, and others, as belonging. Roth's (2016) quotation of Vygotskji's (2005) phrase perhaps best encapsulates this: 'we become ourselves through others' (p. 1021). Those researchers that draw on Gee's (2000) concept of 'affinity identity' (e.g. Deneroff, 2016; Hsu et al., 2017; Richardson, 2019) acknowledged science teacher identity as being socially constructed.

What is missing from this body of research is an integrated approach that draws together the individual aspects, or facets, of identity (e.g. personal, professional, situated, affinity, objective, subjective) as well as an understanding of the social processes that form identity (e.g. competition, balance, dissonance) over time. Without this integrated approach, there is a greater likelihood that researchers will be less able to draw insights from across identity-focused research, with its multiple, separate contexts and diverse foci, with the consequence that there would be the potential for learning to be lost (Rushton & Reiss, 2020). Rushton and Reiss (2020) have argued that the SIA provides both the coherent overview and the nuanced understanding of the process of identity formation and development that the field needs. Common to all studies reviewed is a recognition that identity is both individually and socially informed. For example, researchers have recognised the importance of community (Glass, 2019), field (Melville et al., 2007), subject (Beijaard et al., 2000) and social justice (Marco-Bujosa, McNeill, & Friedman, 2019). Whilst all these aspects involve ideas and experiences that go beyond the self, the integral nature of shared identity and group membership has not previously been considered across such a diverse and extensive body of literature and acknowledged as a fundamental part of identity formation and re-formation. Absent from research considered in this review is an explicit expression of *identification*, which lies at the heart of the social identity approach, namely that group membership shapes an individual's behaviour to the extent that their social identity derived from this group membership is incorporated into their sense of self (Haslam, 2017). It is not enough simply to acknowledge that identities are both individually and socially informed; according to the social identity approach, group membership fundamentally alters the way an individual's identity is constructed and developed and, therefore, how they behave. As part of a wider conceptualisation of science identity (as opposed to science teacher identity), Avraamidou (2019) highlighted the importance of recognition and emotions as core features of identity. Recognition, or how people are recognised by others, Avraamidou (2019) argued, is an 'ineradicable' aspect of the human world and as such requires an understanding of how socio-economic and cultural contexts privilege some identities and diminish and exclude others. Rushton and Reiss (2020) contend that recognition, as part of identification, is a fundamental part of the SIA, where the groups that people identify with are acknowledged and given validity by the recognition of others in a way that shapes their attitudes, beliefs, and values and behaviours. However, Rushton and Reiss (2020) also recognised that exploring the nature and importance of emotions in identity formation is not explicitly outlined in the SIA and this is a potential weakness of this approach when recognising the intersectional nature of science identity formation.

A further way that the social identity approach provides a useful lens is the concept of *interaction*, which develops and galvanises social identities (Haslam, 2017). A gap in the literature identified in this review is the explicit understanding of the role of *interaction* in shaping the extent to which individuals feel part of group membership. For example, Chung-Parsons and Bailey (2019) used the work of Carlone and Johnson (2007) and defined 'science identity' as a combination of performance, competence and recognition, rather understanding science identity as being achieved through shared group membership. The concept of 'affinity', identified as being present in the literature reviewed, could also be helpfully understood as part of *interaction*. For example, the affinity observed in the development of science teachers' philosophies in Hsu et al. (2017) could be understood as teachers developing trust in classroom contexts through shared identities as learners. In contrast, a lack of support from preservice teacher mentors, as observed in Ortega, Correa Molina, and Fuentealba Jara (2014), limits the generation of shared identities between student teachers and mentors. Glass (2019) aligned shared group membership with science teacher identity, in contrast to an individual's science identity, which Glass (2019) suggested is a personal affinity towards science. As Rushton and Reiss (2020) have argued, science identity entails an affinity with both science and scientists, and therefore group membership is a fundamental part of both science teacher identity and science identity.

Relatedly, Seiler (2011) showed that science identity development for some ethnic minority groups was activated and nurtured through experiences and relationships outside of the classroom and school laboratory because the presentation of science within the school context was at odds with their identity. This is consistent with a wider body of research that explores science identity development in young people, where an individual's perception that science is 'not for people like me' contributes to disengagement and disaffection with science (Dawson, 2014). Similarly, the concept of *interactions*, where shared social identity is linked to developing a sense of efficacy, agency and power, could be a very helpful way of understanding how affinity develops and how it might be promoted as part of teacher education and professional development programmes. As Richardson (2019) suggested, affinity can be developed through both face-to-face and online media, which could provide a vital opportunity for teachers who may be isolated subject specialists within schools (due to teacher shortages or geography) to develop shared identities that enhance their agency and effectiveness. Identity development in preservice teachers has been a specific focus of researchers (e.g. Chung-Parsons & Bailey, 2019; Hong et al., 2017; Hsu et al., 2017; Ortega et al., 2014), and these studies recognise the frequently emotionally demanding and time and labour-intensive nature of teacher training and accreditation programmes. Using the concepts of *identification* and *interaction*, the SIA suggests how individuals may in some circumstances move (back) towards different group memberships as a source of positive social identity. For example, a trainee teacher who is changing their career may, during periods of stress and tension, move their shared identity back to their former profession and, during periods when their self-efficacy as a teacher is high, move their shared identity towards teaching.

Building on the work of Hanuscin et al. (2014) and Luehmann (2008), future directions for research that explores the role of groups in teacher identity development could extend our understanding of the ways in which online activities (e.g. blogging), networks and social media platforms (e.g. Twitter) create group membership between those who are physically separate by developing online proximity through these virtual platforms. This research theme could also enable comparisons between identities developed online and offline, allowing researchers to consider the ways in which these different sources of identities may function when enacted in the classroom. In this way, and across the other examples previously described, understanding the role of group membership not only illuminates our considerations of identity development, but may also provide additional perspectives and approaches to enhance teacher well-being, particularly in the context of challenging context of teacher recruitment and retention.

Over the course of Chap. 1 and this chapter a significant amount of literature relating to school science teacher identity has been considered, as well as an overview of the opportunities and challenges relating to school science teacher recruitment and retention. Drawing on these literatures and debates provides a detailed context through which to understand both the formation of science teachers' identities and the contexts which shape this process of identity formation. In Chap. 3, the context for this current study is outlined, through descriptions of the methodological approach and the analytical process. Finally, a detailed understanding of the participants' experiences and insights, and an overview of the themes, are shared.

References

Adams, J. D., & Gupta, P. (2017). Informal science institutions and learning to teach: An examination of identity, agency, and affordances. *Journal of Research in Science Teaching, 54*(1), 121–138.

Akkerman, S. F., & Meijer, P. C. (2011). A dialogical approach to conceptualizing teacher identity. *Teaching and Teacher Education, 27*(2), 308–319.

Antink-Meyer, A., & Brown, R. A. (2017). Second-career science teachers' classroom conceptions of science and engineering practices examined through the lens of their professional histories. *International Journal of Science Education, 39*(11), 1511–1528.

Avraamidou, L. (2014). Studying science teacher identity: Current insights and future research directions. *Studies in Science Education, 50*(2), 145–179.

Avraamidou, L. (2016). *Studying science teacher identity. Theoretical, methodological and empirical explorations.* Rotterdam: Sense.

Avraamidou, L. (2019). Science identity as a landscape of becoming: Rethinking recognition and emotions through an intersectionality lens. *Cultural Studies of Science Education*, 1–23. https://doi.org/10.1007/s11422-019-09954-7

Bang, E. J., & Luft, J. A. (2016). Practices and emerging identities of beginning science teachers in online and offline communities of practice. In L. Avraamidou (Ed.), *Studying science teacher identity: Theoretical, methodological and empirical explorations* (pp. 261–294). Rotterdam: Sense.

Beauchamp, C., & Thomas, L. (2009). Understanding teacher identity: An overview of issues in the literature and implications for teacher education. *Cambridge Journal of Education, 39*(2), 175–189.

Beijaard, D., Meijer, P. C., & Verloop, N. (2004). Reconsidering research on teachers' professional identity. *Teaching and Teacher Education, 20*(2), 107–128.

Beijaard, D., Verloop, N., & Vermunt, J. D. (2000). Teachers' perceptions of professional identity: An exploratory study from a personal knowledge perspective. *Teaching and Teacher Education, 16*(7), 749–764.

Boucher, K. L., & Murphy, M. C. (2017). Why so few? The role of social identity and situational cues in understanding the underrepresentation of women in STEM fields. In K. I. Mavor, M. J. Platow, & B. Bizumic (Eds.), *Self and social identity in educational contexts* (pp. 93–111). Oxford: Routledge.

Bryce, N., Wilmes, S. E., & Bellino, M. (2016). Inquiry identity and science teacher professional development. *Cultural Studies of Science Education, 11*(2), 235–251.

Burke, P. J., & Stets, J. (2009). *Identity theory*. Oxford: Oxford University Press.

Carlone, H. B., & Johnson, A. (2007). Understanding the science experiences of successful women of color: Science identity as an analytic lens. *Journal of Research in Science Teaching: The Official Journal of the National Association for Research in Science Teaching, 44*(8), 1187–1218.

Chi, H. J. (2009). *Development and examination of a model of science teacher identity (STI).* Unpublished doctoral dissertation. Retrieved from http://rave.ohiolink.edu/etdc/view?acc_num=osu1259763038

Christ, O., van Dick, R., Wagner, U., & Stellmacher, J. (2003). When teachers go the extra mile: Foci of organisational identification as determinants of different forms of organisational citizenship behaviour among schoolteachers. *British Journal of Educational Psychology, 73*(3), 329–341.

Chung-Parsons, R., & Bailey, J. M. (2019). The hierarchical (not fluid) nature of preservice secondary science teachers' perceptions of their science teacher identity. *Teaching and Teacher Education, 78*, 39–48.

Cruwys, T., Gaffney, A. M., & Skipper, Y. (2017). Uncertainty in transition: The influence of group cohesion on learning. In K. I. Mavor, M. J. Platow, & B. Bizumic (Eds.), *Self and social identity in educational contexts* (pp. 193–208). Oxford: Routledge.

Dawson, E. (2014). 'Not designed for us': How science museums and science centres socially exclude low-income, minority ethnic groups. *Science Education, 98*(6), 981–1008.

Day, C., Kington, A., Stobart, G., & Sammons, P. (2006). The personal and professional selves of teachers: Stable and unstable identities. *British Educational Research Journal, 32*(4), 601–616.

Deneroff, V. (2016). Professional development in person: Identity and the construction of teaching within a high school science department. *Cultural Studies of Science Education, 11*(2), 213–233.

Dominguez, C. R. C., Viviani, L. M., Cazetta, V., Guridi, V. M., Faht, E. C., Pioker, F. C., & Cubero, J. (2015). Professional choices and teacher identities in the Science Teacher Education Program at EACH/USP. *Cultural Studies of Science Education, 10*(4), 1189–1213.

Dreon, O. (2008). *New science teachers' descriptions of inquiry enactment*. Unpublished doctoral dissertation. Retrieved from https://www.researchgate.net/profile/Oliver_Dreon/publication/253123692_New_science_teachers%27_descriptions_of_inquiry_enactment/links/571e06bc08aed056fa2261bc/New-science-teachers-descriptions-of-inquiry-enactment.pdf

Eick, C. J., & Reed, C. J. (2002). What makes an inquiry-oriented science teacher? The influence of learning histories on student teacher role identity and practice. *Science Education, 86*(3), 401–416.

El Nagdi, M., Leammukda, F., & Roehrig, G. (2018). Developing identities of STEM teachers at emerging STEM schools. *International Journal of STEM Education, 5*, 36. https://doi.org/10.1186/s40594-018-0136-1

Ellemers, N., & Haslam, S. A. (2012). Social identity theory. In P. Van Lange, A. Kruglanski, & T. Higgins (Eds.), *Handbook of theories of social psychology* (pp. 379–398). London: Sage.

Garner, J. K., Hathcock, S., & Kaplan, A. (2016). *Exploring the impact of teacher professional development on a veteran science teacher's professional identity: A case study*. Paper presented at the 2016 annual meeting of the American Educational Research Association, Washington DC.

Garner, J. K., & Kaplan, A. (2019). A complex dynamic systems perspective on teacher learning and identity formation: An instrumental case. *Teachers and Teaching, 25*(1), 7–33.

Gee, J. P. (2000). Identity as an analytic lens for research in education. *Review of Research in Education, 25*, 99–125.

Glass, R. (2019). Science identities in the making. *Cultural Studies in Science Education, 14*(1), 69–76.

Hanuscin, D. L., Cheng, Y. W., Rebello, C., Sinha, S., & Muslu, N. (2014). The affordances of blogging as a practice to support ninth-grade science teachers' identity development as leaders. *Journal of Teacher Education, 65*(3), 207–222.

Haslam, S. A. (2017). The social identity approach to education and learning: Identification, ideation, interaction, influence and ideology. In K. I. Mavor, M. J. Platow, & B. Bizumic (Eds.), *Self and social identity in educational contexts* (pp. 19–52). Oxford: Routledge.

Haslam, S. A., Reicher, S. D., & Platow, M. J. (2011). *The new psychology of leadership: Identity, influence and power*. New York: Psychology Press.

Hong, J., Greene, B., & Lowery, J. (2017). Multiple dimensions of teacher identity development from pre-service to early years of teaching: A longitudinal study. *Journal of Education for Teaching, 43*(1), 84–98.

Hsu, P. L., Reis, G., & Monarrez, A. (2017). Identity discourse in preservice teachers' science learning autobiographies and science teaching philosophies. *Canadian Journal of Science, Mathematics and Technology Education, 17*(3), 179–198.

Izadinia, M. (2013). A review of research on student teachers' professional identity. *British Educational Research Journal, 39*(4), 694–713.

Jetten, J., Iyer, A., & Zhang, A. (2017). The educational experience of students from low socio-economic status background. In K. I. Mavor, M. J. Platow, & B. Bizumic (Eds.), *Self and social identity in educational contexts* (pp. 112–125). Oxford: Routledge.

Jones, S. E., Livingstone, A. G., & Manstead, A. S. R. (2017). Bullying and belonging: Social identity on the playground. In K. I. Mavor, M. J. Platow, & B. Bizumic (Eds.), *Self and social identity in educational contexts* (pp. 70–90). Oxford: Routledge.

Lave, J. (1996). Teaching, as learning, in practice. *Mind, Culture and Activity, 3*(3), 149–164.

Luehmann, A. L. (2008). Using blogging in support of teacher professional identity development: A case study. *The Journal of the Learning Sciences, 17*(3), 287–337.

Luft, J. A. (2009). Beginning secondary science teachers in different induction programmes: The first year of teaching. *International Journal of Science Education, 31*(17), 2355–2384.

Luft, J. A., Firestone, J. B., Wong, S. S., Ortega, I., Adams, K., & Bang, E. (2011). Beginning secondary science teacher induction: A two-year mixed methods study. *Journal of Research in Science Teaching, 48*(10), 1199–1224.

Luft, J. A., & Zhang, C. (2014). The pedagogical content knowledge and beliefs of newly hired secondary science teachers: The first three years. *Educación Química, 25*(3), 325–331.

Manning, A. (2017). *Urban science teachers exploring how their views and experiences can influence decisions to remain in post or not.* Unpublished doctoral dissertation. Retrieved from https://core.ac.uk/download/pdf/141244962.pdf

Marco-Bujosa, L. M., McNeill, K. L., & Friedman, A. A. (2019). Becoming an urban science teacher: How beginning teachers negotiate contradictory school contexts. *Journal of Research in Science Teaching.* https://doi.org/10.1002/tea.21583

Mavor, K. I., Platow, M. J., & Bizumic, B. (Eds.). (2017). *Self and social identity in educational contexts.* Oxford: Routledge.

Mehli, H., & Bungum, B. (2013). A space for learning: How teachers benefit from participating in a professional community of space technology. *Research in Science & Technological Education, 31*(1), 31–48.

Melville, W., Bartley, A., & Fazio, X. (2013). Scaffolding the inquiry continuum and the constitution of identity. *International Journal of Science and Mathematics Education, 11*(5), 1255–1273.

Melville, W., Wallace, J., & Bartley, A. (2007). Individuals and leadership in an Australian secondary science department: A qualitative study. *Journal of Science Education and Technology, 16*(6), 463–472.

Mujtaba, T., & Reiss, M. J. (2013). What sort of girl wants to study physics after the age of 16? Findings from a large-scale UK survey. *International Journal of Science Education, 35*(17), 2979–2998.

Nichols, S. L., Schutz, P. A., Rodgers, K., & Bilica, K. (2017). Early career teachers' emotion and emerging teacher identities. *Teachers and Teaching, 23*(4), 406–421.

Oakes, P. J., Haslam, S. A., & Turner, J. C. (1994). *Stereotyping and social reality.* Oxford: Blackwell.

Ortega, C. M. V., Correa Molina, E., & Fuentealba Jara, A. R. (2014). La práctica del profesor de Ciencias: Significados personales y experiencias de profesores en formación. *Perspectiva Educacional, 54*(1), 17–34.

Reynolds, K. J., Subasic, E., Bromhead, D., & Lee, E. (2017). The school as a group system: School climate, school identity and school outcomes. In K. I. Mavor, M. J. Platow, & B. Bizumic (Eds.), *Self and social identity in educational contexts* (pp. 55–69). Oxford: Routledge.

Richardson, W. D. (2019). *Who will stay? How teacher professional identity influences teacher retention decisions in North Carolina secondary science teachers.* Unpublished doctoral dissertation. Retrieved from http://www.lib.ncsu.edu/resolver/1840.20/36379

Roth, W.-M. (2016). Becoming and belonging: From identity to experience as developmental category in science teaching and teacher education. In L. Avraamidou (Ed.), *Studying science teacher identity: Theoretical, methodological and empirical explorations* (pp. 295–320). Rotterdam: Sense.

Rushton, E. A. C., & Reiss, M. J. (2019). From science teacher to 'teacher scientist': Exploring the experiences of research-active science teachers in the UK. *International Journal of Science Education, 41*(11), 1541–1561.

Rushton, E. A. C., & Reiss, M. J. (2020). Middle and high school science teacher identity considered through the lens of the social identity approach: A systematic review of the literature. *Studies in Science Education.* https://doi.org/10.1080/03057267.2020.1799621

Schachter, E. P., & Rich, Y. (2011). Identity education: A conceptual framework for educational researchers and practitioners. *Educational Psychologist, 46*(4), 222–238.

Seiler, G. (2011). Becoming a science teacher: Moving toward creolized science and an ethic of cosmopolitanism. *Cultural Studies of Science Education, 6*(1), 13–32.

Tajfel, H., & Turner, J. C. (1979). An integrative theory of inter-group conflict. In W. G. Austin & S. Worchel (Eds.), *The social psychology of intergroup relations* (pp. 33–47). Monterey, CA: Brooks-Cole.

Turner, J. C. (1982). Towards a cognitive redefinition of the social group. In H. Tajfel (Ed.), *Social identity and intergroup relations* (pp. 15–40). Paris: Editions de la Maison des Sciences de l'Homme.

Turner, J. C., Hogg, M. A., Oakes, P. J., Reicher, S. D., & Wetherell, M. S. (1987). *Rediscovering the social group: A self-categorization theory*. Cambridge, MA: Blackwell.

van Dick, R., & Wagner, U. (2001). Stress and strain in teaching: A structural equation approach. *British Journal of Educational Psychology, 71*(2), 243–259.

van Dick, R., & Wagner, U. (2002). Social identification among school teachers: Dimensions, foci, and correlates. *European Journal of Work and Organizational Psychology, 11*(2), 129–149.

Varelas, M., House, R., & Wenzel, S. (2005). Beginning teachers immersed into science: Scientist and science teacher identities. *Science Education, 89*(3), 492–516.

Vygotskji, L. S. (2005). *Psychology of human development*. Moscow: Eksmo.

Webb, A. W. (2012). *'Supporting' beginning secondary science teachers through induction: A multi-case study of their meaning making and identities*. Unpublished doctoral dissertation. Available from ProQuest Dissertations (2012. 3525801).

Wenger, E. (1998). *Communities of practice: Learning, meaning, and identity*. Cambridge, UK: Cambridge University Press.

3

Research Approach, Context, Methods and Results

3.1 Research Approach

As discussed in Chaps. 1 and 2, researchers have indicated the importance of investigating science identities in relation to future choices around studying science and science-related careers. Science identities are not merely a reflection of an individual's ability to 'do' science but arise from 'multiple layers of interactions between a student (their identity, background, cultural resources) and science settings' (Archer et al., 2017, p. 742). Avraamidou (2019) highlighted the importance of 'recognition' and 'emotions' as core features of science identity. Recognition, or how people are recognised by others, Avraamidou (2019) argued, is an 'ineradicable' aspect of the human world and as such requires an understanding of how socio-economic and cultural contexts privilege some identities and diminish and exclude others. In the case of school teacher identities Avraamidou (2014) suggested that teacher identity is:

(a) socially constructed and constituted;
(b) dynamic and fluid and constantly being formed and reformed; and

E. A. C. Rushton, *Science Education and Teacher Professional Development*, Palgrave Studies in Alternative Education, https://doi.org/10.1007/978-3-030-64107-8_3

(c) complex and multifaceted, consisting of various sub-identities that are interrelated (p. 164).

Rushton and Reiss (2020) argued that the 'social identity approach' (SIA) provides a lens through which to better understand middle and/or high school science teacher identity. Understanding the role of identity in shaping the experiences of high school science teachers is important and timely given the challenging global context of specialist science teacher recruitment and retention (particularly chemistry and physics specialists) in a number of wealthy countries (Foster, 2018; Marco-Bujosa, McNeill, & Friedman, 2019; O'Doherty & Hartford, 2018; Weldon, 2018).

The use of qualitative research approaches, including collecting and analysing data through interviews with participants, is particularly appropriate when seeking to understand social phenomena, such as the experience of science teachers who are research active. A key strength of qualitative approaches is that they facilitate the formation of new theories (Cohen, Manion, & Morrison, 2018). In recent times, quantitative data, and particularly Randomised Control Trials (RCTs), have been viewed as the 'gold standard' for some 'evidence-based approaches' to education research (Wrigley, 2018). In RCTs, the efficacy of an education intervention is measured through the comparison of results from one group of participants (e.g. school students) who have received the intervention and the second group providing a 'control', and potentially receiving resources or other compensation for their participation at the end of the intervention period. Despite the increasing prevalence of RCTs in education research, it is important to remember the 'real world' nature of education, with all the complexities and social and cultural contexts that shape individuals' identities and decision-making. In order to effect change, it is important first to understand the reasons why things are as they are, to establish not just 'what works' but also *where* it might work, *for whom* and *under what conditions* (Wrigley, 2018). Qualitative data can help to clarify benefits of participation in education initiatives, such as collaborations with research scientists, which might not be immediately apparent in recruitment and/or retention rates.

This study shares the insights of 53 teachers and technicians, or key informants. These key informants were recruited from the study predominantly from the Institute for Research in Schools (IRIS) network and in the following section I provide an overview of IRIS and some contextual information about the research projects in which teachers and technicians have participated.

3.2 Research Context: IRIS—A Network of Research-Active Teachers and Technicians

The Institute for Research in Schools (IRIS) is a UK-based charity launched in March 2016 to develop an approach to school education where research is a key element of STEM (science, technology, engineering and mathematics) learning, offering opportunities for students to work on genuine problems (Parker, Fox, & Rushton, 2018; Rushton & Reiss, 2019). This approach resonates with the concept of what is sometimes termed 'authentic learning'. This term is used with a range of meanings but in the context of school science education it has been argued that practical work is more 'authentic' than much of what goes on in school laboratories when it helps demonstrate or it replicates the sort of work that scientists frequently undertake in modern science, or if it is perceived as having relevance to solving real-life problems (Braund & Reiss, 2006, p. 1378) as when students and teachers are contributing to knowledge by focusing on what is not already known, as part of an inquiry that has value beyond the classroom (Bennett, Dunlop, Knox, Reiss, & Torrance Jenkins, 2018; Lombardi, 2007; Newmann, Marks, & Gamoran, 1996).

Students and teachers and technicians collaborate with active researchers based in universities and industry and IRIS supports schools in building research networks and provides access to data and experimental equipment. This is a social constructivist approach to learning, where students are supported by their peers, teachers and other collaborators, to develop both their understanding of science and to further science itself (Parker et al., 2018). The role of the school teacher and/or technician is to encourage, support and facilitate their students' participation. At the outset, IRIS was not established to provide professional development opportunities for teachers of science (which has been extensively done in England through a range of providers, e.g. the National STEM Learning Centre). However, one of the knock-on effects of IRIS' approach to science education has been to enable science teachers to engage with STEM research. Support and guidance for teachers are provided by IRIS staff and research scientists associated with individual projects through a combination of webinars, training videos, written materials, email groups, school visits and a mentoring system where more experienced teachers support other schools in their geographical area.

Over the last three decades, researchers have documented opportunities and approaches that actively involve young people in practical independent research projects (IRPs) as part of their high school science education (Albone, Collins, & Hill, 1995; Bell, Urhahne, Schanze, & Ploetzner, 2010; Bennett et al., 2018). Research that considers the nature and impact of IRPs in high school students' science education (Bennett et al., 2018) shows that such projects can have considerable benefits for students in terms of engagement and a richer understanding of what it is for a scientist to undertake research. An IRP can be considered as a student-led, extended, open-ended investigation involving practical work, using Millar's (2004) definition of practical work, that is, work that encompasses activities involving students in observing or manipulating the objects and materials they are studying. IRPs are usually undertaken during normal school hours, sometimes supplemented with time in after-school clubs. Occasionally, time is created within schools through 'intensive pull-outs' whereby students are taken off their normal timetable for a period to be dedicated to IRP work. In some countries, IRPs are associated with dedicated out-of-school events such as one- or two-week summer schools and camps.

In a systematic review of IRPs (Bennett et al., 2018) indicated that they are often associated with wider initiatives such as authentic science, problem-based learning and project-based learning. There is considerable variability in the nature of IRP work in relation to such things as the involvement of external partners such as universities and employers, funding and assessment. Most of the published literature on IRPs reviews explores areas such as conceptual understanding, motivation to study science once it is no longer compulsory, attitudes to science and the development of practical skills. Benefits of IRPs are found in relation to the learning of science ideas, affective responses to science, views of pursuing careers involving science and development of a range of skills. Studies focusing on traditionally under-represented groups indicate that such students felt more positive about science as a result of undertaking IRPs. A more recent report of a single study (Dunlop, Knox, Turkenberg-van Diepen, & Bennett, 2019) evaluated open-ended investigative project work for post-16 science students, where such projects give students a certain amount of autonomy but do not necessarily require them to undertake original research. It found that teachers who encouraged their students to engage in such projects believed that they were valuable for much the same reasons reported by teachers who enable their students to undertake IRPs.

IRIS provides an invaluable network of research-active science teachers and an opportunity to expand our relatively limited understanding of the experiences of research-active teachers and technicians, who mentor school student research (Rushton & Reiss, 2019; Walkington & Rushton, 2019). Four projects, *CERN@school, Genome Decoders, MELT* and *Well World*, provide representative examples of the types of engagement and collaborations between school students, teachers and/or technicians and research scientist partners across the wider group of projects experienced by teachers in this research. Indeed, 47 of the 52 key informants had participated in at least one of these four projects.

3.2.1 CERN@school

The IRIS project *CERN@school* lends a radiation detector and software to schools which enables the visualisation of alpha, beta and gamma radiation particles (Parker, Thomas, Rushton, & Hatfield, 2019). Since 2008, *CERN@school* has included more than 300 schools, hundreds of teachers and technicians, and thousands of students. Resource materials and online support has been developed by IRIS, and teachers and technicians have used the technology both to support curriculum activities and provide stimulus for student research, which has included research into solar radiation levels (Furnell, Shenoy, Fox, & Hatfield, 2018). *CERN@school* research projects have led to the development of research projects mentored by teachers and technicians, which have explored radiation in space and large-scale data analysis resulting in co-authored research articles (Furnell et al., 2018; Whyntie & Harrison, 2014, 2015). In these ways, students, teachers and technicians worked collaboratively to produce new understandings and, in some cases, co-publish research outputs.

3.2.2 Genome Decoders

During 2017–2019, teachers, technicians and students from over 60 schools participated in *Genome Decoders*, a project which was supported by scientists at the Parasite Genomics team at the Wellcome Sanger Institute and the WormBase team at EMBL-EBI (European Molecular Biology Laboratory-European Bioinformatics Institute). Students were trained to annotate the genome of the human whipworm (*Trichuris trichiura*) (Rushton & Parker, 2019). The whipworm is a parasite that causes *trichuriasis* (whipworm infection), a neglected tropical disease that affects over 500 million people globally, mainly children living in the poorest parts of the tropics, including Africa, Asia and South America (Barda, Keiser, & Albonico, 2015). Annotating the genome enables the identification of the genes that code for proteins and which are important during infection and the manifestations of the disease, which can lead to identifying treatment and or cure. To annotate a gene, students interpreted the sequencing data generated by *Illumina* and *Pacific Biosciences* sequencing platforms to identify regions of the genome containing genes, and to annotate transcript structures in terms of their constituent exons and introns. To enable school students to contribute to the annotation of the genome, a framework based around the web-based genome annotation editing tool *Apollo* (Lee et al., 2013) was designed. An important aspect of the framework was detailed tagging and tracking of annotations, which enabled work to be allocated, checked and reported on at the level of individual students, and this information was shared with their teachers through the web-based platform. Genome annotation using *Apollo* or similar tools is usually performed by postgraduate and post-doctoral researchers and professional curators and this project required students and teachers to understand and apply complex language and concepts related to genome research. In these ways, students, teachers, and technicians worked collaboratively to contribute to the annotation project.

3.2.3 *Monitoring the Environment, Learning for Tomorrow* (MELT)

MELT is supported by researchers at the Centre for Polar Modelling and Observation, University of Leeds and funded by the UK Space Agency. Launched in early 2018, MELT enables student to research environmental change through (1) the Carbon Footprint Challenge and (2) Earth Observation. The Carbon Footprint Challenge is appropriate for students aged 7–18 years. Students use a carbon calculator to measure the carbon footprint of their school community, develop a plan to reduce their carbon footprint over a defined period of time (e.g. three months) before recalculating the carbon footprint for a second time. Students work in teams to produce a research poster to present to their school communities and through IRIS conferences (Rushton,

Charters, & Reiss, 2019). The Earth Observation strand is appropriate for students aged 14–18 years and many participating students study geography as well as traditional STEM subjects. Students use images from the European Space Agency's Sentinel-1 satellite and Synthetic Aperture Radar data to measure iceberg formation and movement on the Antarctic Peninsula. Over the last 20 years there have been significant iceberg break-off or 'calving' events including at the 'Larsen-B Ice Shelf' in 2002 and the 'Larsen-C Ice Shelf' in 2017. These events may suggest that environmental conditions in the Antarctic Peninsula have changed. Teachers and students share their analysis with scientists and receive feedback and further guidance through electronic-based networking including emails and webinars. Students have presented their research findings at school conferences hosted by IRIS and the Royal Society in 2019.

3.2.4 Well World

The *Well World* research project involved school students (mainly aged 16–18 years, studying biology and/or psychology) designing a range of experiment to explore the initial research question, 'Does biodiversity make us happy?' (Rushton, 2019). As part of these experiments, students identified three areas in their school site and/or local area with different levels of biodiversity and asked participants to walk for the same distance and/or length of time. These areas included an indoor school area (no biodiversity), an outdoor running track (low biodiversity) and the highly biodiverse areas (e.g. woodland) (Rushton, 2019). Participants' physical well-being was measured before and after walking through pulse and blood pressure measurements and their mental well-being was assessed through the shortened Spielberger State Trait Anxiety Index (STAI) questionnaire (Marteau & Bekker, 1992). Students and teachers were initially motivated to research the links between biodiverse areas and health and well-being to raise awareness and appreciation of the value of green spaces within their school community (Rushton, 2019). When developing their study, students considered research that demonstrates the positive role green space can have in promoting physical and mental well-being (Barton & Pretty, 2010; Burgess, Harrison, & Limb, 1988; Pretty,

Peacock, Sellens, & Griffin, 2005; Wood, Hooper, Foster, & Bull, 2017). Green space includes a range of environments including national parks, public parks and domestic gardens (Kaplan, Kaplan, & Ryan, 1998). In the context of increasing global concern about the protection of the environment (including, anthropogenic global warming, deforestation, soil degradation and water and air pollution) (Millennium Ecosystem Assessment, 2006) and rising levels of child and adult physical inactivity and obesity (UK Department of Health, 2009) multi-disciplinary initiatives including Green Gym (Pretty et al., 2007) and Blue Gym (White, Pahl, Wheeler, Fleming, & Depledge, 2016) have sought to reconnect populations to their environments and promote positive health outcomes associated with physical activity in the natural environment. Students presented their research as part of an IRIS student conferences in 2017 and 2018.

Drawing on the experiences of 48 teachers and 5 technicians shared through interviews, I will explore the answers to the following research questions:

- What are the experiences of high school teachers and technicians who are research active?
- What are the professional identities of research-active science teachers and technicians?
- What are the key challenges and opportunities that teachers and technicians experience when they are research active?

3.3 Methods

3.3.1 Key Informants

All 53 key informants recruited for the study were from the IRIS network of teachers and technicians. This included 48 high school teachers (23 female, 25 male) and 5 technicians (5 female). At the time of the study, all teachers taught one or more of the following subjects, biology, chemistry, physics, general science and psychology, at high school level (Table 3.1). All key informants had been working on research projects

Table 3.1 Subjects taught, professional experience and current management roles of key informants

	For teachers only				Teachers and technicians						
	Main subject taught				Number of years' professional experience					PhD	Management role
	Biology	Chemistry	Physics	Psychology	1–5	6–11	12–17	18+	Mean	Yes	Yes
Female	9	4	9	1	9	5	5	8	12.1	7	13
Male	3	4	17	1	2	7	6	10	15.6	8	22
All	*12*	*8*	*26*	*2*	*11*	*12*	*11*	*18*	*13.9*	*15*	*35*

Table 3.2 Geographical location and school type

School type	Non-selective	Selective	Fee-paying
Location of school	England x30 Scotland x5 Australia Norway	England x9	England x5 Scotland Bangkok
Total number of schools	37	9	7

with their students and with teachers and students from other schools for at least six months and were based in schools that had registered with IRIS. Key informants had diverse levels of experience, ranging from newly qualified teachers to those who had been teaching and/or working in schools for over 30 years (Table 3.1). Fifteen key informants hold a PhD in a science subject and 35 were in management roles, including lead technician, subject leader and curriculum leader (Table 3.1). Throughout Chaps. 4, 5, 6, 7 and 8, quotations are included from interviews with Key Informants using pseudonyms. A list of these pseudonyms is provided in Appendix.

Key informants were predominantly based in England (44) and Scotland (6) but were also located in Australia (1), Bangkok (1) and Norway (1) (Table 3.2). Those key informants located in England and Scotland were drawn from a wide geographic range from Cornwall in south-west England to Stirling in central Scotland (Table 3.2). School types included comprehensive schools that do not select by academic ability (37), non-fee-paying schools that select by academic ability prior to entry (9) and fee-paying schools (7) (Table 3.2). This ensured that the study incorporated the experiences of teachers and technicians from a range of educational contexts with different ideologies. The ethnicity of participants was not requested or disclosed.

As Braun, Terry, Gavey and Fenaughty (2009) have described, key informants typically provide 'in-depth experience and knowledge-based perspectives on under-researched topics' (pp. 113–114). The key informants in our research occupied positions 'inside' both the IRIS network and wider research-active communities and, as such, were members of the communities of practice about which they were speaking. This contrasted with my position as a researcher seeking to understand the experience of research-active science teachers and their identity formation. The accounts shared by key informants related to the experiences of other research-active teachers and to their understandings of supporting and mentoring school student research as part of the wider field of extra-curricular activities.

Key informants were recruited during January 2018–December 2019 through a combination of email requests for participants, suggestions of participants from IRIS staff and from teachers who had a pre-existing professional relationship with the author. An interview schedule was prepared with questions in three main sections: background information, including information about the participant's teaching role and research role; the impact of the research on the participant's experience of the subject taught and the experience of teaching and their sense of self. Interviews, each lasting approximately 40–60 mins, were predominantly conducted during visits to the key informants' school to ensure that they were in a comfortable, familiar environment and the interviews were at a time chosen by the participant, carried out during January 2018–December 2019. Some interviews were carried out via telephone and *Skype*, and two key informants elected to share their contributions in written responses to the interview questions provided by the author. At the outset of the interview issues around anonymity were discussed, with key informant contributions shared using pseudonyms and confidentially with participants and provided each key informant with an information sheet, as part of the project's consent process, approved by a University Ethics Committee. Interviews were audio-recorded and transcribed soon after the interview took place. In presented extracts, […] indicates that some text has been removed.

3.4 Analytical Process

Thematic Analysis (TA) is a method for analysing qualitative data that identifies patterned meaning across a dataset. Braun and Clarke's (2006) articulation of the process has been applied to a variety of disciplines and research areas. The technique has recently been further developed as Reflexive Thematic Analysis (RTA) (Braun & Clarke, 2019) and is described as a subjective, organic and reflexive method of data analysis, where researcher subjectivity is understood as a resource, rather than a barrier to knowledge production. In RTA, researchers actively interpret data and create new meaning through systematic phases of research that are iterative and discursive rather than through the rigid application of a coding framework or codebook. These phases include (1) data familiarisation; (2) coding the data set; (3) generation of initial themes; (4) reviewing themes; (5) defining and naming themes and (6) writing up the analytic narrative in the context of the literature (Braun & Clarke, 2006; Clarke, Braun, & Hayfield, 2015). Through this dynamic and reflective processes, researchers generate new patterns of shared meaning founded upon a central concept or understanding (Braun & Clarke, 2019). That is, themes do not passively emerge from the data (Ho, Chiang, & Leung, 2017).

Data familiarisation occurred during the two-year data collection period, through discussions of both interview data with teachers, research colleagues and IRIS staff. After each interview I wrote reflections and commentaries based upon both the interview and drawing on field notes and observations from school visits undertaken to conduct interviews, within a few days of each visit. These notes and reflections enabled me to foreground my own subjectivities, for example, as a former high school geography and general science teacher who had been research active with school students and also as a former employee of IRIS whose role was to evaluate the impact of participation in STEM research projects for students and their teachers and technicians. Through these reflections I documented my responses as both a researcher, teacher and evaluator; my experience as a high school teacher who developed research projects with students undoubtedly informed my understanding of teacher-student interactions within the space of research. In my role as Director of Evaluation for IRIS I worked with students and teachers in a range of ways, including delivering in-person and web-based webinars to support teachers and technicians who were research active, evaluating the impact of students' participation in research conferences and co-authoring research with students (Parker et al., 2018; Rushton et al., 2019) and teachers (Rushton & Batchelder, 2019; Rushton & Robinson, 2019).

Steps two to five of the RTA process began with an analysis of 17 interview transcripts as part of an initial study of the experiences of teachers who are research active with their students (Rushton & Reiss, 2019). During this analysis I looked for descriptions of key informants' experiences in research. For example, I looked at the ways teachers and technicians described their role in the research process, the opportunities and challenges they encountered and the motivations they had to participate in what was predominantly a voluntary role in addition to their central work commitments. Throughout the sequential but also recursive phases of reviewing and refining initial themes and defining and naming themes I sought to generate new understandings of the experiences of school science teachers and technicians who were research active with their students, drawing on the data provided through 53 interviews. Therefore, my analysis was deductive, that is, directed by existing ideas, in this case the theoretical framings from the wider literature outlined in Chaps. 1 and 2, as well as themes identified in the initial analysis of 17 interview transcripts (Rushton & Reiss, 2019). Analysis was also latent, that is, reporting concepts and assumptions underpinning the data and situated in my familiarity with both working with research-active teachers and my own role as a teacher who supported student participation in authentic research whilst at school.

3.5 Overview of Superordinate and Sub-ordinate Themes

Analysis of the interview data resulted in the identification of five superordinate themes, each with sub-ordinate themes that were identified through the iterative analysis of several initial codes (Table 3.3). These five themes, developed from the analysis of interviews with 53 key informants are highly consistent with those found in the initial study of 17

Table 3.3 Superordinate and sub-ordinate themes identified through RTA (Braun & Clarke, 2019)

Superordinate Theme	Sub-ordinate themes
(A) Freedom to teach	(1) Research as a flexible approach to science education which provides intellectual freedom (2) Freedom from external exams and curriculum constraints (3) Variety of teaching and learning methods and approaches
(B) (Re)connection with science and/or research	(1) (Re)connection with science and research through participating in 'discovery' (2) (Re)connection with science and research through engaging in new subject knowledge (3) (Re)connection with science and research through using novel equipment (4) Research projects connecting teachers with their 'roots' as scientists (5) Research projects reconnecting teachers with their prior experiences as scientists
(C) Collaboration	(1) New and different ways of working with students (2) Working with external partners including scientists, university-staff, teachers and students from other schools (3) Establishing and developing collaborative networks
(D) Professional development	(1) Development of specific skills—both practical and interpersonal (2) Increased Recognition of subject and teacher—by colleagues at school of teacher; increased value of subject with students, for some teachers this recognition led to research providing alternative PD routes to traditional routes such as management (3) Teaching and Pedagogy (4) Challenges experienced
(E) Student/societal development through research	(1) Development of students' inquiry skills (2) Development of students' communication skills (3) Provide opportunities for students to develop wider networks and connections, experience of science careers and research (4) Providing solutions to problems that impact the wider world

teachers (Rushton & Reiss, 2019). Each of these five themes is discussed in detail in the subsequent five chapters before a model of teacher scientist identity is proposed in Chap. 9.

References

Albone, E., Collins, N., & Hill, T. (1995). *Scientific research in schools: A compendium of practical experience*. Clifton: Clifton Scientific Trust. https://doi.org/10.1080/09500690802582241

Archer, L., Dawson, E., DeWitt, J., Godec, S., King, H., Mau, A., … Seakins, A. (2017). Killing curiosity? An analysis of celebrated identity performances among teachers and students in nine London secondary science classrooms. *Science Education, 101*(5), 741–764.

Avraamidou, L. (2014). Studying science teacher identity: Current insights and future research directions. *Studies in Science Education, 50*(2), 145–179.

Avraamidou, L. (2019). Science identity as a landscape of becoming: Rethinking recognition and emotions through an intersectionality lens. *Cultural Studies of Science Education*. https://doi.org/10.1007/s11422-019-09954-7

Barda, B. D., Keiser, J., & Albonico, M. (2015). Human *trichuriasis*: Diagnostics update. *Current Tropical Medicine Reports, 2*(4), 201–208. https://doi.org/10.1007/s40475-015-0063-x

Barton, J., & Pretty, J. (2010). What is the best dose of nature and green exercise for improving mental health? A multi-study analysis. *Environmental Science & Technology, 44*(10), 3947–3955. https://doi.org/10.1021/es903183r

Bell, T., Urhahne, D., Schanze, S., & Ploetzner, R. (2010). Collaborative inquiry learning: Models, tools, and challenges. *International Journal of Science Education, 32*(3), 349–377.

Bennett, J., Dunlop, L., Knox, K. J., Reiss, M. J., & Torrance Jenkins, R. (2018). Practical independent research projects in science: A synthesis and evaluation of the evidence of impact on high school students. *International Journal of Science Education, 40*, 1755–1773. https://doi.org/10.1080/09500693.2018.1511936

Braun, V., & Clarke, V. (2006). Using thematic analysis in psychology. *Qualitative Research in Psychology, 3*(2), 77–101.

Braun, V., & Clarke, V. (2019). Reflecting on reflexive thematic analysis. *Qualitative Research in Sport, Exercise and Health, 11*(4), 589–597.

Braun, V., Terry, G., Gavey, N., & Fenaughty, J. (2009). 'Risk' and sexual coercion among gay and bisexual men in Aotearoa/New Zealand–key informant accounts. *Culture, Health & Sexuality, 11*(2), 111–124.

Braund, M., & Reiss, M. (2006). Towards a more authentic science curriculum: The contribution of out-of-school learning. *International Journal of Science Education, 28*, 1373–1388.

Burgess, J., Harrison, C. M., & Limb, M. (1988). People, parks and the urban green: A study of popular meanings and values for open spaces in the city. *Urban Studies, 25*, 455–473. https://doi.org/10.1080/00420988820080631

Clarke, V., Braun, V., & Hayfield, N. (2015). Thematic Analysis. In J. A. Smith (Ed.), *Qualitative psychology: A practical guide to research methods* (pp. 222–248). London: Sage.

Cohen, L., Manion, L., & Morrison, K. (2018). *Research methods in education* (8th ed.). Abingdon: Routledge.

Dunlop, L., Knox, K., Turkenberg-van Diepen, M., & Bennett, J. (2019). *Practical, open-ended and extended investigative projects in science*. Report to The Gatsby Charitable Foundation. Retrieved from http://eprints.whiterose.ac.uk/146914/.

Foster, D. (2018). *Teacher recruitment and retention in England*. House of Commons Briefing Paper No. 7222. Retrieved from https://researchbriefings.parliament.uk/ResearchBriefing/Summary/CBP-7222.

Furnell, W., Shenoy, A., Fox, E., & Hatfield, P. (2018). First results from the LUCID-Timepix spacecraft payload onboard the TechDemoSat-1 satellite in LEO. *Advances in Space Research, 63*(5), 1523–1540. https://doi.org/10.1016/j.asr.2018.10.045

Ho, K. H., Chiang, V. C., & Leung, D. (2017). Hermeneutic phenomenological analysis: The 'possibility' beyond 'actuality' in thematic analysis. *Journal of Advanced Nursing, 73*(7), 1757–1766.

Kaplan, R., Kaplan, S., & Ryan, R. L. (1998). *With people in mind. Design and management of everyday nature*. Washington, DC: Island Press.

Lee, E., Helt, G. A., Reese, J. T., Munoz-Torres, M. C., Childers, C. P., Buels, R. M., … Lewis, S. E. (2013). Web Apollo: A web-based genomic annotation editing platform. *Genome Biology, 14*(8), 1–13.

Lombardi, M. M. (2007). Authentic learning for the 21st century: An overview. *Educause Learning Initiative, 1*, 1–12.

Marco-Bujosa, L. M., McNeill, K. L., & Friedman, A. A. (2019). Becoming an urban science teacher: How beginning teachers negotiate contradictory school contexts. *Journal of Research in Science Teaching*. https://doi.org/10.1002/tea.21583

Marteau, T. M., & Bekker, H. (1992). The development of a six-item short-form of the state scale of the Spielberger State—Trait Anxiety Inventory (STAI). *British Journal of Clinical Psychology, 31*(3), 301–306. https://doi.org/10.1111/j.2044-8260.1992.tb00997.x

Millar, R. (2004). *The role of practical work in the teaching and learning of science.* Paper prepared for the Committee: High School Science Laboratories: Role and Vision, National Academy of Sciences, Washington, DC.

Millennium Ecosystem Assessment. (2006). *Ecosystems and human well-being.* Washington, DC: Island Press.

Newmann, F. M., Marks, H. M., & Gamoran, A. (1996). Authentic pedagogy and student performance. *American Journal of Education, 104*(4), 280–312.

O'Doherty, T., & Hartford, J. (2018). Teacher recruitment: Reflections from Ireland on the current crisis in teacher supply. *European Journal of Teacher Education, 41*(5), 654–669.

Parker, B., Fox, E., & Rushton, E. A. C. (2018). IRIS—Promoting young peoples' participation and attainment in STEM and reigniting teachers' passion for science education. *Impact Vol. 2* Retrieved from https://impact.chartered.college/article/parker-iris-stem-students-teachers-participation-research/.

Parker, B., Thomas, L., Rushton, E. A. C., & Hatfield, P. (2019). Transforming education with the Timepix detector—Ten years of CERN@school. *Radiation Measurements.* https://doi.org/10.1016/j.radmeas.2019.03.008

Pretty, J., Peacock, J., Hine, R., Sellens, M., South, N., & Griffin, M. (2007). Green exercise in the UK countryside: Effects on health and psychological well-being, and implications for policy and planning. *Journal of Environmental Planning and Management, 50*(2), 211–231. https://doi.org/10.1080/09640560601156466

Pretty, J., Peacock, J., Sellens, M., & Griffin, M. (2005). The mental and physical health outcomes of green exercise. *International Journal of Environmental Health Research, 15*(5), 319–337. https://doi.org/10.1080/09603120500155963

Rushton, E. A. C. (2019). Increasing environmental agency through climate change education programmes that enable school students, teachers and technicians to contribute to genuine scientific research. In W. Leal Filho & S. L. Hemstock (Eds.), *Climate change and the role of education* (pp. 503–527). Springer Climate Change Management Series. https://doi.org/10.1007/978-3-030-32898-6_28.

Rushton, E. A. C., & Batchelder, M. (2019). Education for sustainable development through extra-curricular or non-curricular education. *Encyclopaedia of the UN Sustainable Development Goals. Quality Education*. https://doi.org/10.1007/978-3-319-69902-8_19-1

Rushton, E. A. C., Charters, L., & Reiss, M. J. (2019). The experiences of active participation in academic conferences for high school science students. *Research in Science and Technological Education*. https://doi.org/10.1080/02635143.2019.1657395

Rushton, E. A. C., & Parker, B. (2019). Empowering young people to develop STEM careers through active participation in genuine scientific research. In S. E. Hiller & A. Kitsantas (Eds.), *Enhancing STEM motivation through citizen science programs* (pp. 97–127). Hauppage, NY: Nova Science Publishers, Inc..

Rushton, E. A. C., & Reiss, M. J. (2020). Middle and high school science teacher identity considered through the lens of the social identity approach: A systematic review of the literature. *Studies in Science Education*. https://doi.org/10.1080/03057267.2020.1799621

Rushton, E. A. C., & Reiss, M. J. (2019). From science teacher to 'teacher scientist': Exploring the experiences of research-active science teachers in the UK. *International Journal of Science Education, 41*(11), 1541–1561.

Rushton, E. A. C., & Robinson, N. (2019). Bringing research into the chemistry classroom—Perspectives from a researcher and a teacher. In *Education in Chemistry*. London: Royal Society of Chemistry. Retrieved from https://edu.rsc.org/ideas/encourage-your-students-to-research/4010794.article

UK Department of Health. (2009). *Be active and healthy. A plan for getting the nation moving*. London, England. Retrieved from http://www.laterlifetraining.co.uk/wp-content/uploads/2011/12/DoH-Be-Active-Be-Healthy-2009.pdf.

Walkington, H., & Rushton, E. A. C. (2019). Ten salient practices for mentoring student research in schools: new opportunities for teacher professional development. *Higher Education Studies*. https://doi.org/10.5539/hes.v9n4p133

Weldon, P. (2018). Early career teacher attrition in Australia: Evidence, definition, classification and measurement. *Australian Journal of Education, 6*(1), 61–78.

White, M. P., Pahl, S., Wheeler, B. W., Fleming, L. E. F., & Depledge, M. H. (2016). The 'Blue Gym': What can blue space do for you and what can you do for blue space? *Journal of the Marine Biological Association of the United Kingdom, 96*(1), 5–12. https://doi.org/10.1017/S0025315415002209

Whyntie, T., & Harrison, M. A. (2014). Simulation and analysis of the LUCID experiment in the Low Earth Orbit radiation environment. *Journal of Physics: Conference Series, 513*(2), 022038. https://doi.org/10.1088/1742-6596/513/2/022038

Whyntie, T., & Harrison, M. A. (2015). Full simulation of the LUCID experiment in the Low Earth Orbit radiation environment. *Journal of Instrumentation, 10*(03), C03043. https://doi.org/10.1088/1748-0221/10/03/C03043

Wood, L., Hooper, P., Foster, S., & Bull, F. (2017). Public green spaces and positive mental health–investigating the relationship between access, quantity and types of parks and mental wellbeing. *Health & Place, 48*, 63–71. https://doi.org/10.1016/j.healthplace.2017.09.002

Wrigley, T. (2018). The power of 'evidence': Reliable science or a set of blunt tools? *British Educational Research Journal, 44*(3), 359–376.

4

Freedom to Teach

4.1 Introduction

During this chapter, and the following four chapters, five themes (Table 3.3) are explored:

1. Freedom to teach,
2. (Re)connection with science/research,
3. Collaboration,
4. Professional development and
5. Student and societal development through research.

These are explored through an analysis of teachers and technicians' perspectives shared in interviews, as well as a broad consideration of the wider literature relevant to each theme. Although these chapters consider distinct themes, there is inevitably some overlap and elision between the ideas discussed: this reflects the experience of the teachers and technicians interviewed for this research.

E. A. C. Rushton, *Science Education and Teacher Professional Development*, Palgrave Studies in Alternative Education, https://doi.org/10.1007/978-3-030-64107-8_4

4.1.1 Research and Freedoms

When describing the experience of research as part of the weekly work of a teacher and/or technician, 'freedom' was a word that was frequently used. Freedom was connected to (1) the flexibility of research projects as an approach to science education, (2) freedom from the constraints of curricula and external examinations and (3) the variety of teaching methods and approaches that were available when supporting school-based science research projects.

4.1.2 Research as a Flexible Approach to Science Education Which Provides Intellectual Freedom

Teachers described the intellectual freedom that students experienced as part of research projects, with this way of learning science providing a space for students to 'open their minds' and explore complex ideas, driven by *their own* interest:

> Research gives students the opportunity, the freedom, to really get stuck in, to do their own research, their own planning, trials, experiments, and find out what goes wrong, and then go back and do it again, and gain some real ownership that you only get when students genuinely have freedom. (Sally, biology teacher)

> Research projects allow students the freedom to let us know what they think. (Bella, physics teacher)

Teachers highlighted that the intellectual freedom that students experienced during research projects replicated the freedom and discovery that the teachers recognised as an integral part of the discipline of science, which requires students to experiment and think innovatively:

> Research projects allow students to really enter the world of science research and follow it, and make predictions and hypotheses of their own, and are equipped to test those. (Nathan, biology teacher)

> Science is about the freedom of the unknown and of discovery and research projects help students understand that. (Natalie, chemistry teacher)

> Research is science that is not following a script or a recipe, it is real experimental science and thinking outside the box is a big part of research, it is very open and free. (Sarah, technician)

It was found that the intellectual freedom not only provided a 'wonderful' opportunity but also posed many challenges for students who engaged with research, and some teachers described that it was their role as a teacher to provide support so that students could negotiate this experience successfully:

> [In research] students are more sort of searching around for different things and playing around and seeing what types of things they can find, it is that wonderful freedom that you are giving them, you are saying, 'here is some data, what do you want to do with it?' And the trouble is it becomes a bit too free and you, as a teacher, you have to give them some structure. (Mark, physics teacher)

Participation in research projects was not a universally positive experience for students, at least initially, with some teachers identifying students' discomfort in working in such an intellectually open way, where the potential to make mistakes limited their willingness to engage in research:

> The fear of failure that students have when they begin research projects is a significant barrier that can really hold them back, that freedom you get in research can also make students very anxious because they are frightened of making mistakes and the adults role in research projects is to support them to see that making mistakes is actually the thing that develops something fantastic in the project that they are working on. (Sarah, technician)

The 'fear of failure' described by Sarah, and her recognition that adults (including teachers and technicians) have a vital role in providing students with the emotional and intellectual support to overcome this fear is echoed in Mark's recognition of the support that students require from teachers when they enter an intellectual space which is unbounded and limitless. Teachers also experienced a sense of intellectual freedom through working with students in research, 'with research you have the freedom to keep a list of ideas, an endless list of ideas that you keep adding to over time' (Hasan, physics teacher). The description by Sarah, of research providing a way for students to become comfortable with making mistakes, also features in some teachers' experience of research:

> Working in this way, in research projects with the students, it has given me the freedom to try new things and make mistakes, I worry much less if something goes wrong, because research projects have shown me that that we all learn far more from our mistakes than if it all goes smoothly the first time around. (Annie, general science teacher)

Other teachers described how working with students in a practical, 'hands on' way motivated them to start and sustain student research projects:

> Doing research with students came out of not wanting to fill the time I had clawed back [to study] for an MA in Education and put it into something really positive. I love working with students, I love the different ways they approach things and the freedom that the experimental side of science brings to teaching and learning. (Jacinta, physics teacher)

Another teacher described the role of the research day in protecting his time with his students and said that working in research prevented the other demands of teaching from absorbing all his energies:

> The freedom I get with this research day…it keeps you going, it stops the marking and workload from squeezing everything else out. The kids turn up and they want to do research and I can't say no! (Keith, chemistry teacher)

Here, Keith and Jacinta suggest that the day-to-day role of a teacher limits the capacity for teachers to work with students in research and, at the same time, that research projects bring students a freedom to work in a way that is much close to the real experience of science. Teachers frequently and explicitly contrasted the ways in which research projects brought freedom, with the constraints of external exams and curricula, and this aspect of the flexible, free approach to science education is now considered.

4.1.3 Freedom From External Exams and Curriculum Constraints

As described above, one factor motivating teachers to incorporate research projects into their science teaching was to enhance their students' experience of the reality of science with some teachers suggesting that the school science curriculum did not allow students to engage with an authentic experience of science:

> I bring research into the classroom context because I believe that it is the best way for children to learn about what science is—the science syllabus does not make them realise what science is. (Mabel, biology teacher)

Teachers also described the pleasure they had as educators when they saw research projects enabling students to 'shake free of the constriction of an exam' (Elliott, physics teacher) and to develop a desire to learn science that extended beyond the limits of the school curricula and associated assessments:

> Research projects are almost the opposite of school classroom science, by doing both students see…the freedom and challenge of real science, they are now aware of a whole world out there which they have never seen…it helps them look past the exam and think about where they are going with science, there is a motivation that is greater than simply passing a school exam. (Keith, chemistry teacher)

Words that were often used by teachers to describe the science curriculum included 'dry' and 'dull', and teachers contrasted this experience of lesson-based science with the freedom, innovation and inspiration of research projects:

> Research is making me look outwards again and find real life examples…to inspire the kids…to make the teaching less dry and a bit more interesting for the students. (Francis, physics teacher)

Research projects were described by teachers as, 'the highlight of my week' (Mabel, biology teacher) that gave teachers professional satisfaction and enjoyment that brought relief from 'the grind of repetitious curriculum' (Anthony, physics teacher):

> Well, the curriculum is dull…the times that have given me the most pleasure in teaching is when students have made links that have been outside the curriculum and what is in the textbook, and ask those questions that really take you by surprise. (Natalie, chemistry teacher)

Teachers working in Australia, England and Norway made links between recent changes school science curricula in their contexts with a further movement away from, what they described as 'real-life' science, and argued that research projects enabled students to appreciate the wonder of science:

> Unfortunately, both traditional and modern strictly curriculum-based science teaching which is widely practiced in schools nowadays has a big side-effect, namely it fades away both the inherent fascinations of science itself, and the joy of its teaching and learning process at the same time…research is a way to cure this side effect of the current science curriculum. (Hasan, physics teacher)

Teachers also explained how research projects brought clarity to classroom-based practical science which developed students' understanding of the purpose of experimental work:

> I thought that with the change of syllabus that teaching science has lost a lot of the practical element, so when we do practical now it is really short and sharp, and I got the feeling that sometimes the students didn't really understand what they were doing it for. They weren't particularly interested in the results, they just wanted to get it done and out the way and on to the next thing but those who are involved in research club are able to make the links between practicals and science knowledge. (Sally, biology teacher)

In addition to the parallels drawn by some teachers between the 'dull', 'dry' science curriculum and the freedom of research, other teachers described the curriculum as 'safe' and 'secure' and positioned research projects as an 'uncertain' and 'exciting' experience for teachers, which required professional courage:

> Research helps a teacher to be gradually 'brave enough' to move towards the unknown borders of science teaching, something that teachers can be very anxious about, the unknown, and therefore prefer not to do, and so they choose the 'safe and secure' prescriptions of curriculum-based-approaches. (Hasan, physics teacher)
>
> It is very tempting to do chalk and talk as a science teacher, it is tempting and safe so that the students get the right answer, whereas research is much more open, which can be scary because, as a teacher you have to let go of control. (Sophie, biology teacher)

One teacher discussed in detail the ways in which research projects developed their method of teaching, saying that when working on research projects with her students she was able to allow them the time needed to develop their ideas with minimal direction from her. She contrasted this with the teaching method required in classroom science when a large amount of content meant there was not enough time to allow students the freedom to make mistakes and learn from them in classroom experiments:

> My role with the students in the research project is that it is much less directed, I mean obviously in the classroom we try to get them to come up with their own ideas and work things out themselves, but with huge time pressures and curriculum content only increasing you are having to push them in the right direction, but with this…it is a much freer way…you are planning and analysing over a long time, like proper research, whereas in the classroom experiment, it doesn't go right, oh well, this is what should have happened and you move on to the next thing. (Sally, biology teacher)

Another teacher also reflected upon the ways she taught students as part of a research group compared to the classroom context, again highlighting the ways in which research projects require emphasis on problem solving where teachers provide prompts through posing questions rather than learning large amounts of information that students can reproduce:

> With the research group I teach very differently, it is giving some instructions, a prompt, some information and saying, 'off you go!… I started the research projects at lunchtime because I wanted the kids to see what science was about, that it was about evidence, thinking, problem solving, because I don't think it has changed, I think that the research projects have fitted with this approach to teaching more than the school syllabus, which is more about learning extensive amounts of facts and then being about to reproduce them in exam conditions…so instead of saying, look at this, learn this, it is a case of saying, here is the data, figure it out, what would you do with that information? What conclusions would you draw? What would your next steps be? (Mabel, biology teacher)

Some participants acknowledged that this approach would not be one chosen by all science teachers:

> Well, this role doesn't suit everyone, but I think that part of being a scientist is about…just trying something, seeing if it works and if it doesn't try something else, experiment, observe, report, evaluate, that is the scientific method really, and that doesn't suit everyone. (John, physics teacher)

Teachers working in this way recognised that other teachers might not have the opportunity to replicate this in their schools, perhaps limited by timetable constraints but, that others would not have the desire to engage in research projects, 'not all teachers have the space in their timetables or the inclination to take part, I have tried to persuade them!' (Annie). Time constraints are a barrier described by many, and teachers identify that there are times of the year (March–June) and phases of education (e.g. students aged 15–16 years) where there is no time for research activities as the focus was on external examinations (e.g. GCSEs):

> There is so much pressure on students and teachers for great results at GCSE that it is almost impossible to get students the time they need to do this [research] in Years 10 and 11. They want to, they do it in Years 7–9 and then the pressure kicks in and then you hope you can get them back in later years. (Barbara, chemistry teacher)

> If I am honest, I have to say that getting research into the classroom, rather than as an extracurricular activity, is still a challenge, and it is a challenge because we are constrained by the requirements of the examinations. (John, physics teacher)

Other teachers also suggested that lack of time was the significant factor in limiting the amount of research both they and the students could do:

> The greatest challenge with research is time; I wish I had more time to devote to Genomics as a teacher… I would also like to spend more time supporting the students and I wish they had more time to devote to it, but they are spending time on a Friday after school so they really are committed, they just don't have any more time. (Jane, biology teacher)

Here, Jane is describing her work with students aged 16–17 years, demonstrating that in some cases, teachers and students were able to incorporate research into their experience of school science. In contrast to this understanding that students (particularly during certain examination phases) did not have control over their time, there was a limited recognition that it was possible for teachers to 'make time' if they wanted to incorporate research into part of their and their students' experience of teaching and learning science: to be honest, more teachers could do this [research] if they really put their mind to it (Nell, chemistry teacher). Other teachers recognised their ability to take on the research aspect of their professional life was as a direct consequence of having a leadership role in school, and/or because they were the only subject specialist in school, so were able to take decisions about curriculum delivery that meant they could incorporate research as part of classroom teaching rather than simply as an extra-curricular club or activity:

> There are various aspects of how I run it [research], I am trying as much as possible to bring it into the class…rather than just a club so that it is an incentive to do physics, if you have to come into a physics classroom to get hold of things that are as cool as the CERN equipment…and well, I am the only physics teacher in the school and I am the head of department…and that is helpful, maybe that is what gives me the freedom? (Gordon, physics teacher)

Freedom and time available are linked themes since a participant who feels free to do research will feel they have more time available to undertake research compared to a teacher who feels more constrained. In an earlier phase of this research which included interviews with 17 teachers, those who identified with a sense of freedom had been teaching for at least ten years, and some had positions of responsibility; Rushton and Reiss (2019) contended that this may have contributed to their increased sense of freedom to pursue research, compared to those teachers who were in earlier phases of their teaching career. In this larger study, which includes interviews with 53 teachers and technicians, the average amount of teacher experience is relatively high at just under 14 years and two-thirds have a current management role (Table 3.1). However, this study also includes teachers at much earlier stages of their careers, including newly qualified teachers and those with less than five years' experience, which suggests that participating in research projects with students is not something that is limited to teachers in the mid-to-later stages of their careers (Table 3.1). Of the 53 teachers and technicians interviewed as part of this research, 28% held doctorates (Table 3.1). Although this is perhaps higher than might be expected of the general high school science teacher population, a substantial majority of teachers and technicians do not hold a PhD and as such, research projects cannot be considered as restricted to those with prior experience of research in doctoral or post-doctoral contexts.

4.2 Practical Approaches to Research

In their discussions of research projects, some teachers (e.g. Mabel and Sally) referred to a variety of teaching and learning approaches within their research projects that extended and developed their classroom-based experiences. The following section will consider these approaches in more detail in the context of wider research that explores the role of independent research projects (IRPs) in science education.

4.2.1 Variety of Teaching and Learning Methods and Approaches

Broadly speaking, teachers highlighted the variety that working in research projects brought both their, and their students', experience of science:

> The research projects bring variety, different ways of approaching science, for example, a new piece of equipment that is a tool to ask and answer questions, and that variety frees students up to see that there are many different ways to work in science, to be a scientist. (Declan, chemistry teacher)

Research projects provided new ways for students to experience science that were 'real-world, multi-phase and time-consuming' (Hasan, physics teacher). Furthermore, this varied experience of science 'frees students up to see that there are many ways to work in science, to be a scientist' (Dean, biology teacher), enabling young people to generate a multiplicity of imagined futures in science that could promote careers in science, technology, engineering and mathematics (STEM). Wider research has also found that student participation in scientific inquiry can have positive impacts upon attainment and retention in STEM subjects (Trautmann, Shirk, Fee, & Krasny, 2012), and Wieman and Gilbert (2015) suggested that science research projects can increase attainment and retention in school contexts. Mitchell et al. (2017) suggested that inquiry-based learning for both school and undergraduate students increases attainment as this approach to science education can 'promote the retention of material by increasing deep thinking in students' (p. 2). The ability to focus on 'hands-on' practical work was also a key feature of the experience of teachers who supported research projects:

> I came into education as a mature student, I have had quite a varied background myself, and the way I learn things is by doing them and finding out for myself instead of reading from a book. Research is a very practical, hands-on way of learning, and the students get to do much more that compared to the classroom lessons. (Keith, chemistry teacher)

As has been outlined in Chap. 3, key elements of an independent research project (IRP) include an investigation that is student-led, extended, open-ended and involves practical work. In their discussions and reflections on their experiences of research projects with students, teachers focused on the following four elements within the broader framework of research providing variety and a focus on practical science:

1. the role of student autonomy and direction,
2. research groups with a range of ages and experiences,
3. learning beyond the confines of the classroom and
4. open-ended and iterative phases of learning.

4.2.2 Student Autonomy and Student-led Research

Independent research projects are founded upon an approach to learning that is inquiry-based, which has been described as seeking, 'to move beyond a pedagogy of instruction and recall of facts to enhance or produce a pedagogy of engagement and practice' (Edwards, McDonnell, Simpson, & Wilson, 2018, p. 197). Bell, Urhahne, Schanze, and Ploetzner (2010) suggested that in inquiry-based learning, students are supported by their teachers to ask their own research questions, evaluate evidence and explain and disseminate results. Teachers described the ways in which students were able to choose the focus of research, and identify the research questions and methods, following their own interests:

> Students follow their own interests, almost without limits…some projects run for multiple years, driven by the students as they move up the school…it progresses along as a storyline, the whole research is a very rich story which they are writing as they go. (Keith, chemistry teacher)

The practical focus was also a key motivator for student participation, as were the high levels of student autonomy:

> The students that were involved were incredibly motivated from the start, they researched and collected data for themselves, they wanted to something really practical which was great. (Tahira, physics teacher)

One teacher, Mark, reflected that the students he worked with were more autonomous than he had realised at the time, developing strategies to 'manage-up' the adults around them, including their teachers:

> During the student's final presentation it was quite interesting to hear how they had to overcome using me, their teacher, as a resource in research, to use me effectively, so they were definitely leading the work, even if I didn't always realise it! (Mark, physics teacher)

The real-world, tangible nature of practical science, which involved high levels of student choice was described by one teacher as crucial to student engagement:

> With the soil testing…they are able to use the equipment, being able to take a test and then being able to see the results and choosing the test and it is different from just seeing a piece of paper which describes the experiment, when they actually do the experiment themselves…especially when the students know that these soil tests exist and are used every day by scientists in the government and farmers and they are using the tests themselves, this is not a hypothetical example, these are tests that are done by people who grow crops, by people who have livestock, and the students get a kick out of it, as none of it is a construct. (Sophie, biology teacher)

Bonney, Phillips, Ballard, and Enck (2016) suggested that science research projects can provide a useful way of engaging students who might not otherwise be interested in science and scientific inquiry, especially when those projects have an overarching narrative that resonates with the student, for example environmental conservation (Bonney et al., 2016; Butler & MacGregor, 2003). The importance of a compelling narrative in providing the 'hook' that increases student engagement and motivation for scientific inquiry has been found in wider studies of school student research (Rushton & Parker, 2019a, 2019b) and this aspect is explored in greater depth in Chap. 8.

As has been discussed previously in this chapter, teachers highlighted the ways in which independent research projects, and inquiry-based learning is an approach to science education that contrasts with classroom-based science teaching, bringing intellectual freedom, variety and a focus on practical, experimental science with high levels of student autonomy. Teachers also recognised that this is not always a comfortable experience for students, who find the open-ended nature uncomfortable and require support and scaffolding from teachers and technicians in order to thrive. The student-directed nature of the research was identified as particularly difficult by some:

> Sometimes student-led work can be problematic, students hit hurdles and it is the confident ones that will say that they have reached a dead-end and need help, the less confident ones will go quiet and sit on the sidelines. (Mabel, biology teacher)

> Research projects are much more student-led than that curriculum, they do need some prompting because they can sort of fumble around without some starting direction, but the majority of the project is student-led. (Amy, physics teacher)

Here, Mabel and Amy describe differences in student ability and the role of the teacher in identifying when students reach the limit of their current understanding and need support in order to progress. One reason for this variety of ability and experience within research projects is that they are frequently 'vertical' teaching groups, with students from different year groups working on the same project or, working in parallel on different projects and this aspect of teaching and learning in research projects is now considered.

4.2.3 Research Projects and the Role of Vertical Teaching Groups

Mabel described how she intentionally created a research group with a mixture of ages and abilities, working on a variety of projects and that this creates a positive, energetic atmosphere:

> I created a research hub, with students at different ages and stages, different questions, different data sets, all working alongside one another, with a teacher in the room providing guidance where necessary, prompts, resources, overcoming a bump… I really love the buzz, the feel of the space, it is quite open, and at the beginning I thought we are never going to get the aim of the project done, but it does get there, it has to develop. (Mabel, biology teacher)

Sophie also highlighted the benefits of having research teams with a broad age range, with older students working as near-peer mentors for the younger students in multiple smaller groups, with overarching teacher support:

> Having mixed age groups in research teams works really well…the older students guide the younger students without doing it for the younger students, they make them come up with information, they make them actually do the research themselves and just guide them along, so for me it is fabulous because I just give them some direction. (Sophie, biology teacher)

Sarah shared how she 'actively encouraged' mentoring within the research projects, and that mentoring of younger students by older students enabling large numbers (about 50) to participate in a range of projects, supported by only one or two staff members at any one time. In addition to older students mentoring a younger student, Mark describes the way that vertical groups enable students with previous experience of research to 'inspire' students who are new to research. Mark suggests that a successful part of the verticality is that seniority within research can be decoupled from age, and that younger students who have more experience are comfortable to mentor and guide older students with less experience:

> Mixed aged groups are really successful, it is about your position in the research club, not necessarily your age, you will have younger students, say 14 years old, who have more experience and they will happily boss around students three years older than them, because the younger students have experience to share and are helping to develop the older students, which is great to see, and we see friendship groups often cross years, rather than stay within year groups. (Mark, physics teacher)

The opportunity for teachers to work with students in vertical groups not only enables them to support a greater range of projects, with a larger number of students, but it is also a way of working that is not frequently available in the school curricular context which predominantly focuses teaching around year groups. This can provide challenges, with students working at different levels and, especially during the early phases of research, where students required greater support in order to get started:

> At the beginning, what happens is that there are about five or six things going on simultaneously and it looks chaotic and it looks messy, but this is just what it is like during the early stages of research, when we are all figuring out what we are doing, and I need to give as many students as possible support. (Mabel, biology teacher)

Mark, Sarah and Mabel describe how research provides a context where age is not always synonymous with experience or expertise, and where interest and engagement in science are important prerequisites for participation. Although research groups are sometimes delivered across different year groups for pragmatic reasons (staffing and other resource availability), teachers frequently noted the resulting benefits.

4.3 Different Approaches to Learning Through Research: Beyond the Classroom, Play and the 'Maker Mindset'

Teachers discussed approaches to learning that they experienced in research projects that are largely distinct from formal science learning including (1) learning beyond the classroom, (2) play and (3) making and the 'maker mindset'.

4.3.1 Drawing on Learning Contexts Beyond the Classroom

Perhaps unsurprisingly, some biology teachers suggested that 'it is a whole different way learning when you are outside' and emphasised the way that research encouraged them to incorporate authentic data collection in outdoor environments that required them to provide varied support for students:

> It is really nice from my perspective as a biology teacher to have them out and about, and if they are a bit worried about being outside, I try to get the students to interact at the level that leads to their interest a little more so that they get more out of it, because obviously I don't want them sitting at the side, not interacting. (Sophie, biology teacher)

Beyond this preliminary, 'change of scene', teachers recognised that research enabled students to demonstrate their interest in science and their scientific knowledge and understanding in learning environments beyond the classroom, and that this provided a greater range of students to receive recognition for their ideas and contributions:

> I support a lot of informal learning in outdoor contexts outside of school, and through that I am quite used to seeing young children with amazing abilities and skills e.g. problem solving, being adaptable, having resilience, that you just don't see in them in a classroom context, but like outdoor learning, you see these abilities in the research environment, so it is great to bring that out in the kids, so that they get to show what they are interested in. (Keith, chemistry teacher)

In a similar way to Keith, Hasan drew on his experience of teaching and learning in outdoor settings as a metaphor for his experience of teaching in the research context. He described how, during the summer holidays, he worked as a mountain guide supporting children and adult learners in outdoor activity programmes:

> I always tell students that learning in research is like climbing a mountain, when you come down from a mountain you have learnt things that you have never had the chance to learn before, often without realising it, just like periods of research, whether that is reading scientific literature or working in the laboratory. You have to know that in research, like mountain climbing, you have a difficult job, going up is harder than going down, you have to be prepared to fall, you have to trust people around you, you can't do it alone, collective knowledge and effort can bring everyone to new heights. (Hasan, physics teacher)

Here, Hasan is describing the challenges of extended research, which requires students to develop resilience and effective teamwork. This aspect of collaborative working is considered in greater detail in Chap. 6.

4.3.2 Positioning Research as 'Play'

Some teachers described how they took a variety of approaches when they positioned research with students, with some teachers explicitly discussing whether or not they described phases of research, and research with different age groups, as 'play'. Before exploring this aspect of teachers' reflections, a broad outline of the literature relating to play and education provides a useful preface.

Play has long been recognised as central to the development of children (Piaget, 1945; Vygotsky, 1978) and yet, is variously defined and described in the literature. Sheridan (1977) suggests a broad definition of play, where children have eager engagement in pleasurable physical or mental effort to obtain emotional satisfaction. Elements of 'eager engagement', 'pleasurable mental effort' and 'emotional satisfaction' are present in teachers' description of being research active with school students who voluntarily participate, work on challenging open-ended projects, and 'get a kick' out of research. Krasnor and Pepler (1980) argue that an activity constitutes play when five elements are present: (1) voluntary participation, (2) enjoyment, (3) intrinsic motivation, (4) pretence and (5) where process is more important than the outcome. When thinking about student participation in independent research projects, four of these five elements appear to be present in many contexts. For example, students are choosing to join research projects or clubs as part of an extra-curricular activity that takes place during lunchtime or after-school sessions. Teachers working with students comment upon the enjoyment that young people experience working in this way, and that students develop intrinsic motivation working as part of a research process that is decoupled from formal, external assessment and curricular. The element of Krasnor and Pepler's (1980) delineation of play that is not visible in independent research projects is 'pretense'; indeed, teachers describe the environment that students are working in as 'real-life' and 'authentic'.

However, given the contrast between the freedom experienced by teachers and students when they work in research and the relative constraints of external exams and curricula, an element of 'pretense' could be interpreted in enacting science in a way that is so different to that of the 'everyday' classroom experience.

Elements of open-ended activity, spontaneity and free choice are present in more recent understandings of play, including Mathieson (2017), who describes play as 'activity, motivation and emotional response specifically including freedom to choose' (p. 602). As has been discussed earlier in this chapter, the sense of freedom for both teachers and students participating in research was described in detail by teachers as a predominantly positive aspect, but that the freedom that open-ended projects engendered was also challenging for some students, who required more support and scaffolding from their teachers as the adapted to this new way of engaging with science. In the following quotation, Elliott explicitly links notion of play with freedom to explore:

> Through research we are opening up the door to a field and saying, 'hey, go and play in the field, have a go'. That has been a real, real benefit to us as teachers, it is giving the teacher the freedom to let students have a go, play around and see where it takes them. (Elliott, physics teacher)

For some researchers, the concept of fun forms an aspect or feature of play. For example, Sutton-Smith (2011) describes play as a voluntary activity that includes fun, excitement, free-choice and is intrinsically motivated. Likewise, Gajadhar, De Kort, and Ijsselsteijn (2008) incorporate fun within their definition of play, suggesting that play is an activity that has no object other than fun, enjoyment or amusement. Howard (2017) highlights the challenges that various understandings of play can bring, namely, that dictionary definitions of play that include elements of frivolity and fun are at odds with the seriousness that is often apparent in children's play. Even though it is possible to identify elements of play in teacher descriptions of student research from a variety of definitions found in the literature (e.g. Krasnor & Pepler, 1980; Mathieson, 2017), the lens of play may not be seen as appropriate for some adults or children when describing the 'serious' or 'authentic' learning experience that is taking place in research projects. For example, one teacher described how if he used the term 'play' with young adolescent students (13–15 years) this would be off-putting, and students would perceive the activity as 'silly' and immature. However, the teacher also recognised the importance of opportunities for students to 'play around with the equipment' and have an 'exploratory phase' with a 'playful approach' where students were able to experiment without viewing 'mistakes' as problematic:

> If you were to say, 'let's play' the students would say, 'oh no, we don't do that, that is being silly, playing, oh no no, we are too big to play'. So I call it a 'pilot study', the students hear 'pilot' and they associate that with serious work, official research, and so in this phase of 'play' or 'pilot' I might start this by saying, 'do you know what this equipment does? Do you know how to put it together?' (Keith, chemistry teacher)

As noted by Howard (2017), the incorporation of fun into some theorisations of play is perhaps more aligned with dictionary definitions that describe play as light-hearted amusement and, as Keith suggests, this elision is found in students' response to the use of the wordplay in the context of science learning. In Keith's framing of an experimental introductory activity that is playful as a 'pilot study', he positions the students' engagement as 'serious' and 'official', and therefore more valuable, than their perception of play. In his use of language, Keith is seeking to use and describe approaches to research projects that match his students' needs and expectations. In the wider literature, playful learning environments have been defined as those environments which encourage experimentation, and that experimentation is necessary to develop knowledge that can be successfully transferred from one context to another (Hatano & Inagaki, 1986). This focus on experimentation is also present in Keith's description of what happens during the pilot phase of student research projects:

> I invite them to try and find out how thinks work, play around with the equipment, see what happens, we work out 'play' as 'do something with this, try and understand how it works' so they get the benefits of that

> exploratory phase, that playful approach where mistakes are not seen as a problem, just part of this phase of learning. (Keith, chemistry teacher)

In contrast to Keith's explicit policing of his language around play, other teachers chose to use the term play in their descriptions of research with their students not only because they saw play as a value learning opportunity for children 'of any age' but also because their use of the term play reduced students' anxiety and fear about the complexity of research, and increased students' perceptions of their ability and aptitude in the context of a research project:

> I think research can give students the chance to play at any age, to have time to observe and explore and play as a way to reach understanding and through that, the children are more motivated to continue, to find out more, to ask more questions because they are less worried about what is right and wrong, I say to them, 'have a play with this idea or this bit of kit and see where it takes you' and they love that because if you are 'having a play' you can't really get it wrong, you only don't succeed if you don't try. (Annie, physics teacher)

Both Keith and Annie recognise the value of play but have chosen to modify their use of the term play depending on their audience. Within Annie's quotation, there is also the suggestion that a playful approach encourages motivation and this idea is also found in the literature, which suggests that activities that are playful develop intrinsic motivation, serving to increase the chances that an individual endures (Read, MacFarlane, & Casey, 2002) and perseveres through (Vansteenkiste, Simons, Lens, Sheldon, & Deci, 2004) periods of challenge. Over the last decade, approaches to informal science learning have expanded to acknowledge the value of pedogogical framings that are focused on play and fun (e.g. Rushton & King, 2020) and, have also included approaches such as making and tinkering. As these ideas are found in teacher's descriptions of research projects, a brief overview and context to these terms follow.

4.3.3 Research Projects, Making and the 'Maker Mindset'

Making activities are increasingly becoming a more familiar part of young people's informal science learning across Europe (Blikstein & Krannich, 2013) and the USA (Martin, 2015), having been introduced into programmes within science centres, museums, libraries, independent for-profit and non-profit organisations with local, national and global reach (Halverson & Sheridan, 2014). Making activities have been variously defined in the literature. Honey and Kanter (2013) highlight their hands-on nature and their collaborative iterative approach to learning. Blikstein (2013), meanwhile, references the role played by technologies and materials in making endeavours. Calabrese Barton, Tan, and Greenberg' (2017) definition draws attention to elements of collaboration and creativity in the making process. Martin (2015) offers a working definition of making as:

> A class of activities focused on designing, building, modifying and/or repurposing material objects, for playful or useful ends, oriented toward making a 'product' of some sort that can be used, interacted with or demonstrated (p. 31).

Making activities are frequently aligned with a learning approach centred on play and playful learning environments (Calabrese Barton et al., 2017; Martin, 2015). Examples of making activities (also called 'makes') include creating a moving insect or 'jitterbug' from a CD, motor and craft materials (e.g. feathers, pipe cleaners, sequins), building a robot using cardboard or creating vehicles made out of Lego to race or display. Whilst these making activities have a different purpose and output to that of research projects, there are features of the 'maker mindset' that are 'commonly professed' in the maker community (Martin, 2015), which have similarities with aspects of the experience of learning through research as described by teachers. Martin (2015) suggests the four features of a 'maker mindset' are 'playful'—play, fun and interest are at the heart of making; 'asset- and growth-oriented'—makers focus on skills that can be developed, rather than fixed abilities; 'failure-positive'—failure is valued as a part of successful making (as it is common to have several 'failures' or iterations before a make is successful) and 'collaborative'—maker spaces promote a collaborative approach and build communities of making (p. 35). Martin (2015) argues for the educational value of 'experimental play' through making and asserts that 'play, fun and interest are at the heart of making' (p. 25), which is very closely aligned to Hatano and Inagaki's (1986) description of the importance of experimentation in playful learning environments. In Keith and Annie's description of research (see pages 23, 24), they describe playful experimentation as a way of developing understanding in students where the students are interested and motivated and less worried about making mistakes. Sarah and Mark (page 3) highlight how students require support to overcome fears and anxieties that can accompany open-ended, free research, which could be described as becoming 'failure-positive'. Gordon also identifies the iterative nature of research, where ideas are returned to and developed over time, that is consistent with the maker mindset:

> I think that, kind of, iterative process of spending some time designing their research projects, then coming back to them and realizing that they have to strip back their ideas and make it simpler…that sort of process as part of scientific thinking is so important to learn, and hard to do in the classroom curriculum, which is not iterative. (Gordon, physics teacher)

In the following quotation, Elliott explicitly links ideas of play and participation in research projects with developing a different approach to learning:

> We have had a real shift in thinking…being more inquisitive…research is a way of introducing that learning environment where you play around, you sort of tinker with ideas and make things. (Elliott, physics teacher)

Whilst theories of learning in making and the 'maker mindset' are still somewhat emergent, the perspectives of teachers suggest that there are similarities between making and research projects, with opportunities for further understanding teaching and learning in both contexts.

The theme of 'freedom to teach' has covered a diverse area, including teacher and student perceptions of intellectual freedom, particular freedom from external curricula and exams, and freedom to describe and implement research using a variety of approaches to teaching and learning. In the following chapter, I explore the ways in which participation in authentic research projects supports teachers and technicians to (re)connect with the discipline of science, through aspects of subject knowledge and the process of research.

References

Bell, T., Urhahne, D., Schanze, S., & Ploetzner, R. (2010). Collaborative inquiry learning: Models, tools, and challenges. *International Journal of Science Education, 32*(3), 349–377.

Blikstein, P. (2013). Digital fabrication and 'making' in education: The democratization of invention. In J. Walter-Herrmann & C. Büching (Eds.), *FabLabs: Of machines, makers and inventors*. Bielefeld: Transcript Publishers.

Blikstein, P., & Krannich, D. (2013, June). The makers' movement and FabLabs in education: Experiences, technologies, and research. In *Proceedings of the 12th international conference on Interaction Design and Children* (pp. 613–616). ACM.

Bonney, R., Phillips, T. B., Ballard, H. L., & Enck, J. W. (2016). Can citizen science enhance public understanding of science? *Public Understanding of Science, 25*(1), 2–16.

Butler, D. M., & MacGregor, I. D. (2003). GLOBE: Science and education. *Journal of Geoscience Education, 51*(1), 9–20.

Calabrese Barton, A., Tan, E., & Greenberg, D. (2017). The makerspace movement: Sites of possibilities for equitable opportunities to engage underrepresented youth in STEM. *Teachers College Record, 119*(6), 11–44.

Edwards, R., McDonnell, D., Simpson, I., & Wilson, A. (2018). Educational backgrounds, project design and inquiry learning in citizen science. In C. Herodotou, M. Sharples, & E. Scanlon (Eds.), *Citizen inquiry: Synthesising science and inquiry learning* (pp. 195–209). Oxford, UK: Routledge.

Gajadhar, B. J., De Kort, Y. A., & Ijsselsteijn, W. A. (2008). Shared fun is doubled fun: Player enjoyment as a function of social setting. In *International Conference on Fun and Games* (pp. 106–117). Springer, Berlin, Heidelberg.

Halverson, E. R., & Sheridan, K. (2014). The maker movement in education. *Harvard Educational Review, 84*(4), 495–504.

Hatano, G., & Inagaki, K. (1986). Two courses of expertise. In H. Stevenson, H. Assume, & K. Hakuta (Eds.), *Child development and education in Japan* (pp. 262–272). New York, NY: Freeman.

Honey, M., & Kanter, D. (Eds.). (2013). *Design, make, play: Growing the next generation of STEM innovators*. London: Routledge.

Howard, J. (2017). *Mary D. Sheridan's play in early childhood. From birth to six years* (4th ed.). Abingdon: Routledge.

Krasnor, L. R., & Pepler, D. J. (1980). The study of children's play: Some suggested future directions. In K. H. Rubin (Ed.), *New directions for child development: Children's play* (Vol. 9). San Francisco: Jossey-Bass.

Martin, L. (2015). The promise of the maker movement for education. *Journal of Pre-College Engineering Education Research (J-PEER), 5*(1), 30–39.

Mathieson, K. H. (2017). Understanding the importance of play in a child's development. *Journal of Health Visiting, 5*(12), 602–604.

Mitchell, N., Triska, M., Liberatore, A., Ashcroft, L., Weatherill, R., & Longnecker, N. (2017). Benefits and challenges of incorporating citizen science into university education. *PLoS ONE, 12*(11), e0186285. https://doi.org/10.1371/journal.pone.0186285

Piaget, J. (1945). *Play, dreams, and imitation in childhood*. New York, NY: W. W. Norton.

Read, J. C., MacFarlane, S., & Casey, C. (2002, August). Endurability, engagement and expectations: Measuring children's fun. In *Interaction design and children* (Vol. 2, pp. 1–23). Eindhoven: Shaker Publishing.

Rushton, E. A. C., & Parker, B. (2019a). Empowering young people to develop STEM careers through active participation in genuine scientific research. In S. E. Hiller & A. Kitsantas (Eds.), *Enhancing STEM motivation through citizen science programs* (pp. 97–127). Hauppauge, NY: Nova Science Publishers, Inc.

Rushton, E. A. C., & King, H. (2020). Play as a pedagogical vehicle for supporting gender inclusive engagement in informal STEM education. *International Journal of Science Education, Part B*, https://doi.org/10.1080/21548455.2020.1853270.

Rushton, E. A. C., & Parker, B. (2019b). Evaluating the impacts on young people and their teachers who participate in genuine scientific research whilst at school. In S. E. Hiller & A. Kitsantas (Eds.), *Enhancing STEM motivation through citizen science programs* (pp. 349–374). Hauppauge, NY: Nova Science Publishers, Inc.

Rushton, E. A., & Reiss, M. J. (2019). From science teacher to 'teacher scientist': exploring the experiences of research-active science teachers in the UK. *International Journal of Science Education, 41*(11), 1541–1561.

Sheridan, M. (1977). *Spontaneous play in early childhood: From birth to six years* (1st ed.). Windsor: NFER.

Sutton-Smith, B. (2011). The antipathies of play. In A. D. Pellegrini (Ed.), *The Oxford Handbook of the development of play* (pp. 110–118). Oxford University Press.

Trautmann, N. M., Shirk, J., Fee, J., & Krasny, M. E. (2012). Who poses the question? Using citizen science to help K–12 teachers meet the mandate for inquiry. In J. L. Dickinson & R. Bonney (Eds.), *Citizen science: Public collaboration in environmental research* (pp. 179–190). Ithaca, NY: Cornell University Press.

Vansteenkiste, M., Simons, J., Lens, W., Sheldon, K. M., & Deci, E. L. (2004). Motivating learning, performance, and persistence: The synergistic effects of intrinsic goal contents and autonomy-supportive contexts. *Journal of Personality and Social Psychology, 87*(2), 246.

Vygotsky, L. S. (1978). *Mind in society*. Cambridge, MA: Harvard University Press.

Wieman, C., & Gilbert, S. (2015). Taking a scientific approach to science education, Part I—Research. *Microbe, 10*(4), 152–156.

5

(Re)connection with Science/Research

5.1 Introduction

Teachers described their (re)connection to science and research, in terms of not only the science subject(s) they taught but also a more fundamental connection to scientific inquiry. For some teachers and technicians, this (re)connection enhanced their sense of being a scientist, and this featured in the experiences of those with and without a PhD in a STEM field. Teachers outlined the different ways in which this process of (re) connection happened including (1) (Re)connection with science and research through participating in 'discovery', (2) (Re)connection with science and research through engaging in new subject knowledge, (3) (Re) connection with science and research through using novel equipment, (4) Research projects connecting teachers with their 'roots' as scientists and (5) Research projects reconnecting teachers with their prior experiences as scientists. Each of these aspects is now described in detail.

E. A. C. Rushton, *Science Education and Teacher Professional Development*, Palgrave Studies in Alternative Education, https://doi.org/10.1007/978-3-030-64107-8_5

5.1.1 (Re)connection with Science and Research Through Participating in 'Discovery'

Teachers frequently used the word 'discovery' when describing their experiences of research projects. For some teachers, this sense of discovery was something that they felt was missing from their work as a schoolteacher, and that research projects had reintroduced this element to their experience of science:

> After leaving research to become a teacher I sometimes hanker after those moments of discovery at the cutting edge, [research projects] have allowed me to rediscover these moments alongside young people. (Nathan, biology teacher)

> I have found myself, that being involved in projects, it enthuses you, being engaged in research, asking questions again, being alongside people in a journey of genuine discovery and challenge, finding ways around problems. (Keith, chemistry teacher)

> Research projects are about discovering things and sharing what we discover, it is having a vision and taking that vision forward, this is what science is all about and it is so enthusing for teachers to be able to engage with physics, with science in this way. (Dean, biology teacher)

Being part of discovery through research enabled some teachers to feel less isolated and removed from science and to connect with other like-minded teachers. One teacher, working as a Director of Science across a group of schools that included three high schools and five primary schools, described how research projects supported both specialists and non-specialists to positively engage with science and discovery:

> In my team, I have some teachers with a research background, but most don't and in primary schools you don't always have a science specialist leading science, but with this variety of background research projects help all of these teachers get involved in research, and they get enthused because they are part of discovery where everybody is trying to find the answer, and

research projects help them see that this is what science is about. (Barbara, chemistry teacher)

Other teachers also linked their experience of novel research to increased enthusiasm and specifically linked this to active involvement in 'discovery' or being 'experimental':

Getting involved in projects has actually really helped me a lot as a teacher, because it has enabled me to do what I am really passionate about because I am able to work in the environmental context and actually be testing and doing the experiment. (Sophie, biology teacher)

The research projects have kept me going, kept me enthused, maybe that is the researcher in me, the scientist in me, I am constantly curious. (Bailey, technician)

You have got to yourself, see the joy in not understanding stuff, the questioning mind that can accept that at times you are not always going to immediately know the answer and to enjoy the challenge in that. It is about being happy to be experimental, where we really don't know the answer to the questions, that has got to be a part of who you are. (Tony, biology teacher)

Teachers recognised that their participation in research projects enabled them to experience 'cutting-edge' research that provided them with the opportunity to learn about new areas and developments in science:

If you're doing these projects, you're looking at new techniques that you might not be aware of, you get to hear about all of this amazing science…doing these projects, it just opens your mind to science. (Jonny, chemistry teacher)

Opportunities to engage with new developments not only generated renewed enthusiasm in teachers but also increased some teachers' confidence in their abilities as both teachers and scientists:

> Through this experience I have developed a really tangible link with research myself…this research project has given me an opportunity to refresh my teaching, I don't want to get stale and this is a wonderful way of not getting stale, whilst actually contributing new data to research science. (Jane, biology teacher)

Another teacher, Madeleine, described research projects as an iterative process of learning alongside her students, where both teachers and pupils were engaged in periods of challenge that required persistence and perseverance and ultimately developed confidence:

> Through this project my own confidence in my abilities as a scientist have grown because I have learnt new skills and I have been put back in that place of not knowing and feeling discomfort, which is where you grow…with the first group of students I said, 'I can't find anything and you can't find anything so how can we solve that? How should we do it differently?' And so the next week we had to agree to scrap what we had done and have another try, which is such an important learning experience for students and teachers, and so unusual for student to watch an adult learn and struggle through something difficult and achieve. (Madeleine, general science teacher)

Teachers describe the broad experience of being part of 'discovery', 'inquiry' and 'experiments' as developing their enthusiasm, enhancing their skills and increasing their self-efficacy as teachers and scientists. As well as this overarching aspect, teachers also identified more specific elements of their experience of research projects as part of a process of (re) connection. The first of which is engaging in subject knowledge that is new to the teacher.

5.1.2 (Re)connection with Science and Research Through Engaging in New Subject Knowledge

Research projects required teachers to engage with topics and subject content that were unfamiliar, and teachers spoke positively about this opportunity to learn new aspects of science: Anthony (physics teacher) said, 'teachers have to be learning all the time, it is not just the students who are doing the learning, so research, it keeps me on my toes, it keeps me active and learning.' An example of a project that required teachers to develop new subject knowledge is *Genome Decoders* (for a summary of this project, see Chap. 3). *Genome Decoders* focuses on genome annotation and the development and impacts of neglected tropical diseases (NTDs). Both teachers with and without postgraduate STEM qualifications relished the opportunity to engage with new areas of their subject specialism, 'I've loved everything I have learnt myself about annotating a genome' (Peter, biology teacher). When discussing a different biology-focused project, Madeleine (general science teacher) said, 'the research project is a great way to link yourself with science…I have learnt lots of things about pollen and bees and wider ecology that I didn't know before, it has been a steep learning curve.' Even when projects such as *Genome Decoders* required teachers to spend their own time learning new material, teachers and technicians saw this as valuable and was a process that connected them with research:

> It took me quite a long time to learn all of the materials myself…This was a big time investment but it also made me feel that I was part of this research, by learning all of this new information about genetics and how to use the new software. (Melanie, technician)

As well as those teachers participating in biology-focused research projects, subject knowledge development was also a feature of the experience of teachers who were involved in physics projects such as *CERN@school* (see Chap. 3). One physics teacher called Michelle described how her involvement in *CERN@school* required her to develop her own subject knowledge:

> Being involved in research projects, meant also ensuring that I was constantly addressing my own subject knowledge and getting to grips with newer science, that was at the forefront of what we know and that is really exciting, that is what it is to be a physicist I think. (Michelle, physics teacher)

Michelle suggests that developing subject knowledge as part of research projects is to be welcomed and explicitly links this to what it is to be a scientist. This (re)connection with being a scientist through new subject knowledge is also described by specialist physics teachers Elliott and Jacinta:

> So, as a physics teacher, part of my motivation is finding things out about the world, and to be able to find something out that is new, through research with students, that is for me what being a scientist is all about. (Elliott, physics teacher)

> I am really passionate about physics, and I have always tried to keep my knowledge up to date because I love the subject and through research projects like *CERN@school*, you are really keeping up to date with current advances in physics research—it provides inspiration for me as a physics teacher but also for me to engage as a physicist, and I think that gives me a professional edge. (Jacinta, physics teacher)

Francis also described how this approach to developing subject knowledge is possible for teachers who are non-specialists, and associates this with 'doing real science':

> I think, as a non-physics specialist who is teaching physics, the research project includes lots of areas that I haven't studied before, so I really enjoy getting to learn this with the students, and it does feel like I am doing real science, as opposed to just standing up and regurgitating science information to students. (Francis, physics teacher)

Furthermore, Michelle describes how she took this experience of research projects beyond her experience as an individual and sought to implement this across her school department to enable staff development:

> A core aim in transforming the science department was skilling up each individual member of staff, so that they are subject experts and through the research projects had a real, contemporary context for what they are teaching. This means that staff are able to include physics research say, at the cutting edge of what we currently know. That brings real confidence,

because teachers develop their own area of subject expertise. (Michelle, physics teacher)

Across subjects (including biology, chemistry and physics) when exploring experiences of non-specialist and specialists at an individual and departmental levels, research projects provide teachers with the opportunity to (re)connect with science and/or research through developing new subject knowledge. An alternative way that teachers described (re)connecting with science and/or research was through the use of new equipment, and this is now considered.

5.1.3 (Re)connection with Science and Research Through Using Novel Equipment

As part of projects such as *Genome Decoders* and *MELT* (see Chap. 3), teachers and students used new software and data sources, including satellite images, which they reported that they 'loved' using and felt 'part of science'. Over the course of the decade-long *CERN@school* project (see Chap. 3), schools were loaned 'kits' including a laptop and an MX-10 radiation detector that used the Timepix technology. These kits enabled teachers and students to detect and visualise ionising radiation, showing young people what they would previously only have heard as clicks on a Geiger Müller (GM) tube (Parker, Thomas, Rushton and Hatfield, 2019). Student engagement through *CERN@school* has, over time, inspired multiple student-research projects (Parker et al., 2019) which have resulted in scientific publications (Hatfield, 2010; Hatfield et al., 2019; Whyntie and Harrison, 2014, 2015). For many teachers, accessing this type of equipment engendered excitement, that they were able to work with 'new software and cutting-edge technology, usually reserved for the engineers at CERN' (Arthur, physics teacher). Gordon (physics teacher) said, 'we have been lent the detector and that has enabled us to do physics that we simply could not do otherwise.' Teachers themselves described the difference that teaching radiation with access to *CERN@school* resources, compared with the traditional GM tube:

> The *CERN@school* detector is much more than a Geiger-Müller tube, it is a sophisticated bit of kit that not only detects the radiation but it lets you see the different radiation particles and it can give you spatial coordinates for them and detects the energy that each particle hits the chip, this is kit that you just don't get normally in schools, and you can simply ask students, and yourself, 'what do you want to find out?' That is such an exciting and open question that this equipment allows you to ask. (Tahira, physics teacher)

The ability to use the detector to ask open research questions, undertake real investigations, and access and contribute to large-scale data sets were advantages identified by teachers for teachers, and these features provided inspiration and promoted active engagement in science research for teachers and students:

> Using the detector brought real physics into my classroom and allowed me and my students to do experimental work that was hugely engaging and would have been impossible without this equipment. I very much hope that all schools across the country have the chance to teach radiation with the use of the detector…the opportunity to carry out research has enthused both students and me as a their teacher to be part of the conversation of science rather than just listening to science. (Gordon, physics teacher)

> I try and bring the new equipment into the classroom as much as possible and show the students that we have borrowed this from CERN, that this is the same technology that is on the ISS, and that they can measure background information using it, it is so great for me as a teacher to be able to provide them with the equipment that enables them to ask genuine scientific questions where we don't know the answer, and to have that opportunity to explore that as a teacher. (Arthur, physics teacher)

As well as specialist physics teachers, non-specialist physics teachers also described the value of the *CERN@school* equipment as a way to promote open inquiry and experimentation, where the teacher has an active role in science:

> Research projects allow you to approach science in a different way, you can have a new piece of equipment, a detector that allows you to ask questions, a huge data set that is the prompt to ask questions. This puts the teacher in the position of being scientist, and a scientist who through tools and data can encourage their fellow scientists', the students', scientific questions. I am not a physicist, I am a chemist, but I can get behind supporting the students to think as scientists, whatever the subject content. (Declan, chemistry teacher)

One teacher reflected upon how the experience of research through the *CERN@school* project helped her position her science teaching so that students saw physics as an area of ongoing discovery, and something they could contribute to:

> I think research projects have had a huge influence on my teaching, up until that point science had mainly lessons of male achievements and discoveries, but when we started physics research projects it was about going to CERN, and using equipment in school that is used by male and female scientists at CERN. At CERN the students saw men and women working in ground-breaking experiments and using the equipment at school was an invitation to participate, they understood that there is more work that needs to be done by the students, now, physics is full of unanswered questions that the students can help answer. (Bella, physics teacher)

Teachers working as part of research projects, which included access to novel equipment and developing subject knowledge, also identified how this experience encouraged them to incorporate more practical work as part of their teaching and learning with students:

> Research projects have been an amazing opportunity to extend myself into the teaching I have always wanted to do, I love the hands on, I love the kids doing practicals, and getting more involved and actually being able to see the fruits of their labours…to see how the seedlings are growing, that is real, practical science. (Sophie, biology teacher)

One teacher highlighted how research projects supported them to implement practical work that modelled the reality of the work of scientist and researcher:

> As a psychology teacher, one of my biggest concerns is that students understand the importance of ethics, and research projects allow me to model that in a real world situation, to work with them as a researcher does, developing consent forms and consider the health and safety implications of research as a psychologist rather than in the abstract as a psychology teacher might. (Stephen, psychology teacher)

Here, Stephen explicitly links his experience of research projects with students to exemplify the work and role of a psychologist. This is also found in the experience of teachers who used software as part of *Genome Decoders* and a detector as part of *CERN@school*.

During this chapter, teachers have described how they moved through the spaces of subject teacher, researcher and scientist as part of a process of (re)connection with science and/or research. In an earlier phase of this research, Rushton and Reiss (2019) considered the experiences of 17 research-active teachers, 6 of whom had achieved a doctorate in a STEM subject. Rushton and Reiss (2019) suggest that those teachers with a PhD may have had a different experience of research projects with students than those teachers who did not (yet) have experience of doctoral research. Rushton and Reiss (2019) describe how teachers with PhDs draw on their prior experiences as researchers in their work with students, and implicit within this narration of teachers' previous research experience is that their identity as a scientist and researcher was established during their doctoral research. In contrast, those teachers without this prior experience of doctoral research explicitly link the development of their identity as researchers and scientists to their work with students. This current study provides an opportunity to examine the experiences of 52 teachers and technicians, where 36 participants do not have a PhD in a STEM subject and 15 participants are educated to doctoral level, with one further person holding a doctorate in STEM Education. This larger participant pool enables a more detailed exploration of research active teachers' identity development and (re)connection with science and research.

5.1.4 Research Projects Connecting Teachers with Their 'Roots' as Scientists

Those teachers without experience of research beyond their undergraduate degrees described how research projects enabled them to connect with that sense of being a scientist that they had developed whilst a university student. For example, Jane (biology teacher) said, 'this project has connected me with my roots as a scientist', whilst Annie (biology teacher) said, 'research projects have had a huge impact on me, I have been a teacher in the same school for decades and this has reminded me I am a scientist.' Sally reflected that her experience of research has fundamentally reconnected her with research and with her identity as a scientist:

> After this experience of research, I would be far more likely to describe myself as a scientist …which is in some ways surprising as I have taught science to students for over 20 years, but it is that connection with current research and academics and being part of the understanding of new science that I think has made me feel far more likely to describe myself as a scientist. (Sally, biology teacher)

A biology teacher involved in *Genome Decoders* described in detail the way in which her participation in the project shaped her professional identity over time, moving towards that of a scientist as well as a teacher of science:

> When I first began teaching, I saw myself as a scientist who taught young people science, but over the last few years that sense of being a scientist had begun to fade. Working on the *Genome Decoders* project has connected me with cutting edge genetics research, and this has enabled me to see myself as a professional who is a teacher, but also a scientist. This experience has helped me realise that I have something to contribute to the science world as well as the teaching profession. (Jane, biology teacher)

Both Natalie, a chemistry teacher with a doctorate, and Sally, a biology teacher without doctoral research experience, contend that teachers do not need a background in research in order to implement research projects with students, if teachers are given support:

> I really hope that there would be a pathway for teachers to be involved in research, even if they haven't got experience in research science, like me, as once I was given support, I found I was more than able…research has given me an enthusiasm, and self-belief that I didn't realise that I had, maybe I had it all along but I lost some of it on the way. (Sally, biology teacher)

> I don't think you need a research background to do research projects with students, if we can move it away from research being an extra science club and making it more about CPD (continuing professional development) for teachers, as a science teacher I would say that I have a responsibility to keep up with subject knowledge and new equipment and technology so research projects are a vehicle for doing that. (Natalie, chemistry teacher)

Natalie highlights that research projects can provide a model for teacher professional development that supports developing subject knowledge which incorporates novel research-related equipment and technology. This is similar to the description outlined by Michelle (physics teacher) earlier in this chapter, where she implemented *CERN@school* across her department as an approach to staff development. In contrast Mark (physics teacher), who developed a second career as a teacher after more than two decades as a nurse, firmly stated that having participated in *CERN@school*, he did not see himself as a scientist, rather that he was 'very much a teacher of science…somebody who tries to get people the knowledge to be able to become a scientist.' However, at a later point during the interview, as part of a discussion about his experiences of research projects with his students, Mark said:

> I have never really been involved in scientific research before, so it is kind of interesting from my viewpoint, being part, in a small way of a scientific project; if I am completely honest, I am doing science for the first time in my life, as opposed to teaching science. (Mark, physics teacher)

The perspectives of Sally and Jane (biology teachers) and Mark (physics teacher) highlight that implementing student research projects, in terms of prior research qualifications and/or experience, is broadly open to all staff: experience of scientific research at postgraduate and beyond is not a pre-requisite for participation. Furthermore, teachers who have not got this prior experience explicitly recommend this as an optional pathway for all staff.

5.1.5 Research Projects Reconnecting Teachers with Their Prior Experiences as Scientists

Within the group of teachers and technicians interviewed as part of this current study, 15 teachers had undertaken research at the doctoral level and beyond. For many of these teachers, student research projects provided them with a chance to draw on that prior professional experience, and share it with their students:

> Working in research with students has allowed me to develop a relationship with them at a different level because I am able to share with them my past world as a research scientist and just seeing them tantalised by that, it gives me a lot of pleasure. (Natalie, chemistry teacher)

Some teachers with doctorates in STEM subjects were involved in school-based research projects that loosely drew on their prior expertise, and they reflect that this supported their work as a teacher, by keeping them refreshed and enthused. Arthur who implemented the *MELT* project (see Chap. 3) with his students said:

> My background before I went in to teaching was as an earth observation scientist…using satellite imagery in its early days…now I'm trying to get students to think about how they can use imagery to think about what we could do with satellite data in the Polar Regions…it benefits me because it keeps me active in the research side of things so, you don't get as stale, perhaps, if you're constantly having to do electrical circuits…if you've got this ability to drop in and do some research with the students it helps to refresh you, keep you active, develop enthusiasm. (Arthur, physics teacher)

Rose, who previously completed doctoral research which explored rice photosynthesis, described how working in research projects enabled her to do the work of a scientist and that reconnected her with her identity as a scientist:

> You need to be active in research in some way I feel to be a scientist, and that could simply be reading the literature and keep up with developments but it is so much more meaningful when you are making that contribution to science. (Rose, physics teacher)

In contrast, other teachers with a STEM doctorate were explicitly clear that their support of student research had no connection with their previous research experiences and did not link this with a sense of identity as a scientist and/or researcher:

> I did a PhD in Zoology, but my involvement in school research was nothing to do with my former research background, my involvement is about exiting young people to do biology. (Peter, biology teacher)

Rose and Natalie reflected upon how their experience of research projects with students enabled them to move between the spaces of scientist, science teacher and researcher and suggest that this movement is multi-directional and is formed over time, shaped by their interactions with research and student researchers. Rose says:

> Before running research projects in school, I would have described myself as a scientist who teaches physics in a school, but not an active researcher anymore. Maybe now that has shifted a little, I mean, research is about moving science forward, so maybe I am more of a physicist in school than a physics teacher? It is something that is evolving, I think. (Rose, physics teacher)

Natalie is less certain that school-based research projects support her identity as a 'research scientist' but that she still retains the 'mindset' of a researcher, and that it is this mindset that she wants to develop in her students:

> I used to be a research scientist, now as a teacher…I guess that is part of me still…I don't know if I would say that is my label, but I still have that mindset and I guess that is what I am trying to engender in the students, through research projects, that research mindset, to encourage them to have fun solving problems with science, so I guess I am still a research scientist, the research bit is where I have fun, it keeps me enthused. (Natalie, chemistry teacher)

This notion of a research 'mindset' also appears in other teacher reflections on their role as a researcher or scientist. One teacher with a PhD in Education was explicit that he was not a scientist or a researcher but that through his work with students, he had 'developed the attitude of a research scientist':

> I have never as a teacher engaged directly in scientific research, I have only supported by students as they engage in research. As an Editor of a student research journal, I have developed the attitude of a research scientist, but I do not see myself as a scientist. (James, physics teacher)

Bella described how in her view, research projects provided a way for teachers, technicians and students to inhabit the 'mindset' of a scientist and that research projects enable teachers to be both teachers and scientists:

> I have never been locked into a mindset of 'I am a science teacher and that is the limit of what I can do and what I can offer.' I have always seen myself as a scientist and a teacher and research projects are a way of sharing that mindset with other science teachers. (Bella, physics teacher)

Within the experiences of the 53 teachers and technicians interviewed for this research, there is variability in the ways in which research projects supported their (re)connection with science and research. For some teachers, the experience of school-based research projects provided an opportunity to develop a sense of their professional identity, possibly for the first time that included that of scientist and researcher. Identity development is multi-directional, changes over time and does not require prior experience as a researcher. The following section considers how teachers' experiences explored in this chapter might be understood in the context of wider literature around inquiry in (science) teaching and learning.

5.1.6 Discovery, Vitality and Renewal in the Context of Inquiry

For some teachers, the aspects of discovery, engaging with new subject content, using new equipment and incorporating more and/or different practical elements developed their sense of (re)connection with science and/or research. This aspect of discovery in the context of 'hands-on', practical science is often described in the wider literature as teachers who implement 'inquiry' strategies as part of teaching and learning or who are 'inquiry-oriented' teachers. There is a substantive literature concerning the term inquiry in the context of science education (some of which has been considered in Chap. 2) with a range of perspectives of what inquiry means in practice. Anderson (2002) highlights that inquiry can relate to (1) scientific inquiry, (2) inquiry learning and (3) inquiry teaching. In scientific inquiry, students develop an understanding of how science functions and how scientists work. Inquiry learning involves students actively building understanding through experiences of practical science, and inquiry teaching uses strategies to promote both scientific inquiry and inquiry learning in the classroom. Melville, Bartley and Fazio (2013) describe teaching and learning of inquiry as being on a continuum, with varying degrees or level of openness to inquiry depending on, 'the teacher-supplied structure, the existence of a solution to the question, the complexity of the activity, or the amount of information that is provided to the student' (p. 1257). Rezba, Auldridge and Rhea (1999) delineate these levels as moving from the least complex 'confirmation' through to 'structured', 'guided' and 'open' inquiry. Teachers interviewed as part of this study were involved in research projects with students that predominantly fall into the category of 'open' inquiry, where they were engaging with research questions where the answers were not yet known. In this context, both the students and their teachers (and technicians) can be positioned as people who are contributing to a wider intellectual endeavour where they are 'capable investigators' (Harris and Ballard, 2018, p. 36).

Many of the studies that have considered the development of science teacher identity in the context of inquiry have focused on preservice teachers (Bryce, Wilmes and Bellino, 2016; Dreon, 2008; Eick and Reed, 2002). As has been discussed in Chap. 2, preservice teachers can experience anxiety when enacting inquiry pedagogy (Dreon, 2008) and benefit from support in terms of scaffolding (Melville et al., 2013), online mentoring (Bang and Luft, 2016) and coaching (Bryce et al., 2016). The USA-based Scientist in the Classroom Partnership (SCP) programme is a substantive example of research that considers inquiry learning beyond a sole focus on preservice teachers (Ufnar, Bolger and Shepherd, 2017; Ufnar, Kuner and Shepherd, 2012; Ufnar and Shepherd, 2019). SCP is a decade-long professional development programme involving 74 high school science teachers who were paired with a STEM graduate student or postdoctoral fellow to co-teach with an inquiry focus one day per week during an academic year (Ufnar and Shepherd, 2019). Features of the SCP programme include gains in discipline and pedagogical content knowledge, inquiry strategies and teacher renewal (Ufnar and Shepherd, 2019). For example, teachers described their increased confidence in learning and implementing inquiry and hands-on strategies as one of the most valuable outcomes of their participation in the SCP programme (Ufnar and Shepherd, 2019). Teachers linked their increased confidence to having the support of a subject expert and found it possible to integrate more inquiry learning into their teaching both during and after the programme (Ufnar and Shepherd, 2019). Teachers also described the importance of gaining greater subject-specific knowledge as part of the SCP programme, a finding which Ufnar and Shepherd (2019) have linked to wider studies that acknowledge the importance of supporting subject knowledge development as a core feature of effective science teacher professional development (Banilower, Heck and Weiss, 2006; Garret, Porter, Desimone, Birman and Yoon, 2001). The experiences of teachers involved in this study also highlight the ways in which teacher engagement with student research supported teachers' own subject knowledge development through exposure to both new material and novel equipment. Teacher reflections on their use of novel, research-grade equipment as part of their engagement with research is an aspect that is, as far as can be determined from the literature, a distinct feature elucidated from this current study. This may be in part due to the nature of the equipment associated with physics-based research project, including the MX-10 radiation detector. However, to the author's knowledge, studies that explore teachers' use of novel equipment as part of inquiry focused teacher professional development and/or school-based citizen science programmes have yet to significantly feature in the literature.

The SCP model of weekly co-teaching by teachers and STEM graduates is distinct from that of the teachers interviewed as part of this research, where teachers were predominantly supported remotely, via email and webinar and when included, face-to-face encounters were limited to perhaps once or twice during the life of the project. However, as Bang and Luft (2016) have demonstrated, online-based subject-specific mentoring can effectively support teachers' agency. In this study, when teachers describe elements of 'discovery', they frequently link this with increased professional enthusiasm and vitality that prevents their teaching from becoming 'stale'. This reinvigoration in the context of inquiry is also visible in the SCP programme, with teachers reporting their increased love of science and describing the SCP programme as 'the most stimulating years of teaching an old-dog new tricks!' (Ufnar and Shepherd, 2019, p. 654). Neither the SCP programme nor the research projects that provided the context for this current study were designed with teacher recruitment and retention in mind and yet, teachers in both studies describe the profound impact that these experiences of inquiry have for their job satisfaction and professional development. As has been described in Chap. 3 and elsewhere (Rushton and Parker, 2019), evaluations of citizen science programmes have predominantly focused on outcomes for students including measuring changes in students' scientific literacy; increasing students' motivation for and engagement with scientific inquiry; increasing student attainment and retention in STEM subjects and promoting positive attitudes towards science-based careers. When the role of the teacher has been considered, this has largely explored the ways in which teachers work alongside students, ranging from gatekeepers and facilitators to roles as co-investigators (Rushton and Parker, 2019). This current study demonstrates the affordances that teacher participation in research projects have regarding (re)connection with science and/or research that are linked with renewal. Future evaluations of school-based citizen science programmes might fruitfully incorporate a more detailed examination of the experiences of teachers and technicians. In the following chapter, I consider the ways in which teachers and technicians experience supporting school student research as a collaborative process which involves a range of external partners and stakeholders including scientists, university-based staff and teachers and students from others schools.

References

Anderson, R. D. (2002). Reforming science teaching: What research says about inquiry. *Journal of Science Teacher Education, 13*(1), 1–12.

Bang, E. J., & Luft, J. A. (2016). Practices and emerging identities of beginning science teachers in online and offline communities of practice. In L. Avraamidou (Ed.), *Studying science teacher identity: Theoretical, methodological and empirical explorations* (pp. 261–294). Rotterdam; Boston; Taipei: Sense Publishers.

Banilower, E. R., Heck, D. J., & Weiss, I. R. (2006). Can professional development make the vision of the standards a reality? The impact of the national science foundation's local systemic change through teacher enhancement initiative. *Journal of Research in Science Teaching, 44*(3), 375–395.

Bryce, N., Wilmes, S. E., & Bellino, M. (2016). Inquiry identity and science teacher professional development. *Cultural Studies of Science Education, 11*(2), 235–251.

Dreon, O. (2008). *New science teachers' descriptions of inquiry enactment.* Doctoral dissertation, Pennsylvania State University. Retrieved from https://www.researchgate.net/profile/Oliver_Dreon/publication/253123692_New_science_teachers%27_descriptions_of_inquiry_enactment/links/571e06bc08aed056fa2261bc/New-science-teachers-descriptions-of-inquiry-enactment.pdf.

Eick, C. J., & Reed, C. J. (2002). What makes an inquiry oriented science teacher? The influence of learning histories on student teacher role identity and practice. *Science Education, 86*(3), 401–416.

Garret, M. S., Porter, A. C., Desimone, L., Birman, B. F., & Yoon, K. S. (2001). What makes professional development effective? Results from a national sample of teachers. *American Educational Research Journal, 38*(4), 915–945.

Harris, E., & Ballard, H. (2018). Real science in the palm of your hand. *Science and Children, 55*(8), 31–37.
Hatfield, P. (2010). Using line intensity ratios to determine the geometry of plasma in stars via their apparent areas. *High Energy Density Physics, 6*(3), 301–304.
Hatfield, P., Furnell, W., Shenoy, A., Fox, E., Parker, B., Thomas, L., & Rushton, E. A. C. (2019). IRIS opens pupils' eyes to real space research. *Astronomy and Geophysics, 60*(1), 1.22–1.24.
Melville, W., Bartley, A., & Fazio, X. (2013). Scaffolding the inquiry continuum and the constitution of identity. *International Journal of Science and Mathematics Education, 11*(5), 1255–1273.
Parker, B., Thomas, L., Rushton, E. A. C., & Hatfield, P. (2019). Transforming education with the Timepix detector—Ten years of CERN@school. *Radiation Measurements*. https://doi.org/10.1016/j.radmeas.2019.03.008
Rezba, R. J., Auldridge, T., & Rhea, L. (1999). *Teaching & learning the basic science skills*. Retrieved from www.pen.k12.va.us/VDOE/instruction/TLBSSGuide.doc.
Rushton, E. A. C., & Parker, B. (2019). Empowering young people to develop STEM careers through active participation in genuine scientific research. In S. E. Hiller & A. Kitsantas (Eds.), *Enhancing STEM motivation through citizen science programs* (pp. 97–127). Hauppage, NY: Nova Science Publishers, Inc..
Rushton, E. A. C., & Reiss, M. J. (2019). From science teacher to 'teacher scientist': Exploring the experiences of research-active science teachers in the UK. *International Journal of Science Education, 41*(11), 1541–1561.
Ufnar, J. A., Bolger, M., & Shepherd, V. L. (2017). A retrospective study of a scientist in the classroom partnership program. *Journal of Higher Education Outreach and Engagement (TEST), 21*(3), 69–96.
Ufnar, J. A., Kuner, S., & Shepherd, V. L. (2012). Moving beyond GK–12. *CBE—Life Sciences Education, 11*(3), 239–247.
Ufnar, J. A., & Shepherd, V. L. (2019). The scientist in the classroom partnership program: An innovative teacher professional development model. *Professional Development in Education, 45*(4), 642–658.
Whyntie, T., & Harrison, M. A. (2014). Simulation and analysis of the LUCID experiment in the Low Earth Orbit radiation environment. *Journal of Physics: Conference Series, 513*(2), 022038. Bristol, UK: IOP Publishing.

Whyntie, T., & Harrison, M. A. (2015). Full simulation of the LUCID experiment in the Low Earth Orbit radiation environment. *Journal of Instrumentation, 10*(03), C03043.

6

Collaboration

6.1 Introduction

The term 'collaboration' is necessarily a broad one, and teachers and technicians who shared their experiences for this study described collaboration in three distinct ways:

1. collaboration as a new and different way of working with students,
2. working with external partners including scientists, university staff and teachers and students from other schools and
3. establishing and developing collaborative networks.

Collaboration was recognised by many teachers and technicians as a key feature of what it is to undertake research and an aspect that was a positive experience that permeated across the school and wider STEM community:

> Students are part of a network or research collaboration that builds year on year, and that their work will continue through the work of other groups…that is how research works, that is not how school works and it is the collaboration that I really love. (Keith, chemistry teacher)

E. A. C. Rushton, *Science Education and Teacher Professional Development*, Palgrave Studies in Alternative Education, https://doi.org/10.1007/978-3-030-64107-8_6

> In my experience there are so many opportunities for collaborations across STEM…this has so many benefits for teachers because they are part of a wider community that is interested in science…you need to have the imagination that science is important to all of us and that we can build cooperative, collaborative work through science, together. (John, physics teacher)

The starting point for much discussion around collaboration was the recognition from teachers and technicians that research projects provided opportunities to work in new and different ways with students. These experiences provide an opportunity to better understand the role of a teacher and/or technician in the context of research projects from their perspective. For example, how do teachers describe their role? How does this role compare to other aspects of their professional practice? What roles do students have in these learning environments? These questions and others are now explored, beginning by considering the ways in which teachers and technicians work with students as part of research projects.

6.2 New and Different Ways of Working with Students: The Role of Teachers and Technicians

Teachers described how both the students and they themselves took on different roles within the learning environment of research projects:

> Research sessions with students have been one of the most enjoyable elements of my teaching role, mainly because I am spending time with young enthusiastic individuals with a real passion for science… it is a very different relationship to the one I would have in lessons with students. (Joseph, physics teacher)

Keith and Nathan identify that in research projects students work more independently, and their role in research projects is more peripheral and responsive to students' needs:

> I work with students in a very different way in the research group, we are working together sometimes, I give them a prompt, and idea and they go with it…other times I am completely in the background. (Keith, chemistry teacher)

> It doesn't take long before you have a group of students all working independently, I mean, I can be sitting in the background sending emails, doing menial teacher jobs whilst they are solving world crises all around me. (Nathan, biology teacher)

Teachers frequently described their role as distinct to their classroom-based teaching role, with Alan (physics teacher) encapsulating the difference as him being 'less a source of knowledge, more of a sounding board.' A variety of terms were used to describe this different role, including advisor, coach, facilitator, guide, mentor and team captain. Those who described themselves as an 'advisor' often linked their role to that of providing 'supervision', linked to broader theoretical frameworks such as 'the scientific method', as opposed to more specific, detailed support:

> I was kind of their [the students] day to day advisor, …I suppose my biggest role was…more about the scientific method, so that this is how you do an investigation, these are the kind of things to think about, this is what a control variable is, that sort of thing. (Dominic, physics teacher)

> I am kind of the students' day to day advisor, my kind of role was to drive the project forward, keeping them on track, giving advice and guidance and we got technical support from external partners. (Robert, chemistry teacher)

For those teachers who described themselves as 'coaches', they highlighted how their role was to provide research projects with an element of 'grounding' so that the work remained 'on track':

> My role is very much a coach, and I don't see it as anything other than that, and perhaps part of the team, and perhaps part of the learning, and I am sometimes the brake, sometimes it is my role to ensure that their enthusiasm doesn't send them running off the tracks. (Mark, physics teacher)

> I think that my biggest job, my role as a coach is to keep the students a little bit grounded, I love the forwarded thinking, but keeping the students a little bit grounded in what we're doing at the moment…I also had to learn to step back, the students had to learn to accept that, they are so used to people telling them what to do now, what to do next. (Sophie, biology teacher)

Teachers who described themselves as 'facilitators' suggested that their role was to support and encourage students to take ownership of the project rather than to 'tell them what to do.' This required teachers to 'take a step back' and sometimes to explicitly articulate this to students, to enable a shift from teacher-led to student-led work:

> I had to tell the students, I am here to facilitate this project, but not to run it…that probably took them the most to get used to, actually being able to say, 'Okay, so can I do this?', and I have to say to them, 'Yes, if that works with this', and they are like, 'Oh, okay, great.' And then they would run with it. (Bethany, physics teacher)

> I am almost more of a facilitator than anything, I guide what they do…I help them to think outside of the box…giving them high expectations of what they can achieve…We discuss the options and then they need to make some choices and decisions about the direction of the project, but I don't spoon feed them, that is not my role in the project. (Sophie, biology teacher)

> I've had to stand back. I've become more of a facilitator, and watching, and probing, and asking questions, just those timely questions to try and make students think. (Ellen, general science teacher)

Other teachers who described themselves as facilitators outlined more practical support that they gave students, for example, brokering contacts, providing logistical help or sourcing equipment, especially for those students who were less confident in seeking out external support. Relatedly, teachers and technicians who described their role in terms of 'guidance' or that of a 'guide' often focused on providing learning environments that were 'safe' for students through direct supervision of research activities, as Sarah (technician) said, 'we are there to supervise the students, to keep them safe but that is it. I am not telling them what to do.' Dominic (physics teacher) also said, 'I was always supervising them when they were doing the experiments from a health and safety perspective.' Teachers also provided oversight during the planning phase rather than direct supervision: 'I provide guidance, I needed to source equipment and to make sure they had plans that were practical and safe' (Amy, physics teacher). In contrast, Melanie (technician) described her role as that of a mentor who provided both the encouragement and an element of 'quality control':

> My role is more of a mentor, I provide encouragement and support, I help with background research…they are a lot more computer savvy than me, but I am helping with quality control, suggesting that perhaps they don't just look at the first page of Google search research results, helping them identify useful search terms. (Melanie, technician)

Technicians frequently outlined how their role was already different from that of a class teacher, both in the nature of their relationships with students and in the practicalities of their position within and beyond the classroom:

> Technicians often have more flexible time, more opportunity to be available when students are…technicians are not in a direct teaching relationship…technicians don't have to get through a curriculum or set learning objectives, for me, technicians simply have the remit of nurturing an enthusiasm for science and to work with students to develop that. (Sarah, technician)

Jane, a biology teacher, described her role with the students as that of an older sibling, with her main role to encourage students to continue when they encountered challenge:

> My role is to be a supportive 'big sister', going around the group and encouraging the students to have a go, saying to them, 'you know lots, you know more than you think?', encouraging students to work it out together and stick with it. (Jane, biology teacher)

Teachers sometimes vividly described how, during research projects, they found themselves in areas of science that went beyond their own experience and subject knowledge. Teachers persisted within this challenging, unbounded space and, as a part of the initial phases of the research project, explicitly invite students to learn something new and demanding with them as equal partners:

> I showed them a screen shot of the software and said to them, 'If that terrifies you, don't worry, it terrifies me, I haven't a clue what to do or how this works, but we can learn together.' (Peter, biology teacher)

> At the beginning I am often completely out of my depth with the science…I have no idea where we are going to end up; I know where we're going to start', so, I just catch up and read up and encourage them to keep going when none of us has a clue what might happen. (Keith, chemistry teacher)

Some participants who describe themselves as advisors, coaches, facilitators, guides and mentors suggest that this role was established from the outset with the implication that this role is consistent throughout their experience of research. However, teachers and technicians also described how their role evolved during the project, from one that was a leadership role and had a didactic approach, to one where leadership and other roles were devolved to students:

> It is quite interesting to see how students work as teams and initially look to me as the team captain…for the first few months it is all about me saying, 'have you done this? Have you done that?'…but in time…there will be certain people who would start to take on the leadership role and…then my role evolves from team captain to more of a cheer leader…for example, there was a student who took control of the data and it was so fun to see him tell the students off for not putting the data entry in clearly and see him tell the others what to do and how they needed to do it and for me to be able to step back and allow them to lead this. (Natalie, chemistry teacher)

Teachers described how the learning environment of the research group had a different 'dynamic' compared to that of classroom-based teaching and learning. For example, through research projects, teachers were able to gain a greater understanding of the 'mentality' of students that enabled them to develop a better 'rapport':

> You get to build a rapport with the students, you develop a professional relationship that supports both their academic development but also their personal attributes like confidence and perseverance and flexibility in thinking. (Rose, physics teacher)

Teachers also suggest that students are aware of, and value, the different learning environment provided through research projects:

> The students like the fact that the research projects involve small groups, that they work in-depth with each other and me over a longer period of time and they get a bit funny if I invite someone else, there is definitely a group dynamic. (Tahira, physics teacher)

These different ways of working for students, and in particular the roles of the student in this learning environment, are now considered.

6.3 New and Different Ways of Working with Students: The Students' Role

Teachers described the research project sessions as having a, 'relaxed, tutorial-like atmosphere, with students working together in twos or threes' (Jane, biology teacher). Within these descriptions, teachers and technicians shared different ways that students worked together in the research environment; for example, some research projects involved students from a single year group, 'the research group of students is gender balanced, and they are all in Year ten, but there is a big range of abilities and personalities' (Clare, chemistry teacher). As has been described in Chap. 4, research projects enable students from across different year groups to work together in 'vertical' cohorts. For some teachers this near-peer mentoring was a central focus of the research project, and teachers saw this as a way of cascading information and inspiration about research projects to increase student participation:

> My main goal for the research project is to produce STEM ambassadors for the school that can enthuse the lower school and recruit them…I want them to share what they have done and then that should inspire others to take part. (Declan, chemistry teacher)

Other teachers used the vertical structure of research projects to provide developmental opportunities for students by giving them different audiences to present to and engage with their research:

> Year nine research project students have presented to younger students and sixth formers…it was quite challenging for them to present to older students, but it helped them get their thoughts in order, to receive feedback on what they had done and the sixth formers were able to help the students think about further questions but the research students also made the sixth formers think, in some ways the Year nine students were showing the sixth formers how to be better scientists. (Mark, physics teacher)

For other participants, the peer-to-peer or near-peer mentoring was a positive consequence of running the research project with students, but not a direct goal or aim. However, teachers did note that research projects promoted an approach to learning that was 'less individualistic', that students were motivated to work together in smaller groups within the setting of a larger group activity and that this was a way of working that was not usually open to them:

> Students learn together, I have a group of students that started in Year seven and have continued through into Year eight and brought in other students, and there is an element that it is this group dynamic that is holding these students together when it gets difficult because they all have different skills and they all share the same research aim…This is almost the opposite to traditional education where you are judged on your own results…maybe drama or fieldwork is different…but not really in experi-

> ments in traditional science and that is not how research works, where you are inspired by your collaborators. (Keith, chemistry teacher)

There are also examples of how mentoring can extend beyond the limits of the school setting, for example, school student alumni have been invited by their teachers to speak to new cohorts of students engaged in research projects, and share their experiences of research and of university science:

> What is really nice, because we have been running research projects for a few years now, we have got students coming back to me who are studying for their degrees, some even postgraduate degrees, they come back to school and they tell the students about their research they are doing and the school students can share their ideas and build that sense of a network. (Nathan, biology teacher)

Other teachers encourage mentoring by enabling their current research students to collaborate with local primary schools:

> We build in an ambassador element to the research project, we expect older students to be ambassadors and go into our local associate primary schools and support them to run a science investigation, we want the older students to share the skills they have learnt through research projects and share those with the younger students. (Dean, biology teacher)

In contrast to many of the descriptions of near-peer and peer-to-peer mentoring, Ralph (physics teacher) recognised the challenges associated with mentoring within the setting of a research project, emphasising the importance of appropriately matching mentors and mentees:

> It is hard sometimes with the mentoring of younger students, the brightest students that would make good mentors in some ways can be a bit off-putting for younger or less-experienced students, they will see these students as super star and this can be a bit overwhelming for the mentee…they may not see that they could be as capable as the mentors. (Ralph, physics teacher)

The students' role as a mentor within research projects is recognised as valuable to both the mentor and the mentee, and a different, more collaborative way of working compared with the more 'individualistic' classroom-based approach to learning. Teacher and technicians also identify the different roles that they inhabit within research projects, drawing on a range of terms including advisor, coach, facilitator, guide and mentor. These terms are used broadly and somewhat interchangeably: however, this use of alternative terms to describe their role seems to be motivated by a desire to distinguish a way of working with students in research groups compared with that within a classroom and curriculum setting. Key elements of this different way of working that promote collaboration include teachers and technicians (1) 'stepping back' and facilitating student decision-making and (2) providing encouragement and support based upon student need. This current research highlights collaboration through new and different ways of working between teachers, technicians and students within the context of the school community, and this is a novel and distinct perspective. Other research, that considers the place of teacher-engagement in STEM research as professional development, has recognised the importance of collaboration; however, to date, researchers have focused on collaborations between groups within school communities (e.g. teachers, technicians, students) and the wider scientific community (Mehli & Bungum, 2013; Ufnar, Bolger, & Shepherd, 2017; Ufnar, Kuner, & Shepherd, 2012; Ufnar & Shepherd, 2019; Varelas, House, & Wenzel, 2005). In the following section, the ways in which teachers and technicians work with external partners through research projects are explored.

6.4 Collaboration Through Working with External Partners

Consistent with other research, including the 'Scientist in the Classroom Partnership' (Ufnar et al., 2012, 2017; Ufnar & Shepherd, 2019), the opportunity to work and collaborate with external partners as part of their experience of research projects was identified by teachers and technicians as an extremely positive aspect:

> I really feel like by doing research projects I have loads of collaborators, students, universities, industry, the parents, they all have different skills and experiences. (Keith, chemistry teacher)

> Through research I am connected to a lot of bright, enthusiastic people who have really great ideas and they are sort of the same as me in that they are all interested in research, so just being able to share in that kind of environment and pinging ideas with all of those people is just great. (Rose, physics teacher)

Collaboration with external partners enabled teachers to develop more opportunities for their students in a variety of ways. For example, Keith (chemistry teacher) shared how his network of 'university researchers and industry people' enabled him to connect with individuals who have access to equipment and chemicals necessary to provide projects for students: 'I just have to go out there and beg, borrow and steal! But I do find that people are willing to help you as a school science teacher.' Other teachers highlighted how collaborations with external partners enabled them to provide further learning opportunities for their students:

> Through this research project we have created more opportunities for the students to meet professionals in different aspects of the project…we organised a trip for students to visit some local beekeepers and learn from them, get their knowledge and expertise of bee populations and they got to do an environmental survey of the area around the hive. (Madeleine, general science teacher)

> We have brought experts in to school because I am not a specialist but the experts can guide us and that helps us, it means I can support a group of super-keen students, we can do that because we have brought that expertise in, rather than just had an initial try at bringing research projects in and then fallen at the first hurdle. (Declan, chemistry teacher)

For Sally, her relationship with an external academic partner, established through the research project, was 'integral' to the success of the project, and enabled Sally to support undergraduate scientists alongside her school students, which further developed her research network:

> Our academic partner provides so much reassurance...the research experience would not have been anywhere near what it has turned out to be had it not had that integral link with an academic...I have felt able to support young undergraduate scientists, their development, alongside my school students which I have really enjoyed. (Sally, biology teacher)

Rose (physics teacher) and Tina (general science teacher) also described the important role that external partners and wider research networks have in providing support:

> Research networks provide you with a further level of support, a group of people that understand what you are trying to achieve and they can work with you, share ideas, tips, contacts, experience. (Rose, physics teacher)

> One thing that has been really important for me has been the involvement of the STEM ambassadors because I am not a specialist physicist, I have a background in biology, and the STEM ambassadors, they have the subject knowledge and the equipment and the background that I don't have...so I think having a lead person from, whether it is from business or somebody that is appropriate for whatever your project is, it adds an extra dimension, and it is a, sort of, go to person as well, in an expert role, which has been really good. (Tina, general science teacher)

Teachers and technicians described the ways in which external partners give authenticity to school-based research projects, and that this is valued by staff and students alike, 'having the external contact with scientists gives the project some authenticity and validity, students are motivated to do the next step' (Ralph, physics teacher). Anthony (physics teacher) also highlights the validation that external partners bring for student engagement in research projects:

> If we can bring external researchers into the school it serves two things, it has that outside link but it also reinforces what we, as teacher, are saying, so that if you have got an academic researcher standing up and supporting what we as teachers are saying, the students can see that, it is almost enhancing and validating the knowledge and expertise of the teaching staff. (Anthony, physics teacher)

As well as the validation that external partners bring, teachers also highlighted the enthusiasm and motivating quality that these collaborations generate for students and teachers:

> One of the things that is so important for the students is that they get to meet other scientists…that is so valuable for them, after those sessions they are buzzing and they are so excited and motivated to carry on, it is so important for them to have that interaction with people they see as real scientists, people they can talk to and ask questions and get feedback from, it expands their understanding of science and what scientists do. (Madeleine, general science teacher)

> Although I am a physics specialist, I am not an astrophysics specialist and I have almost no background in programming, so having your students take on something where they work with external partners and go beyond what you are capable of helping them with, it is just inspiring. (Michelle, physics teacher)

> Working as a team is so much part of it…students in different year groups having to learn to work together…the contact with the academic staff, that helps both staff and students learn more and develop…the feedback and support really keeps us all motivated. (Marina, biology teacher)

Keith (chemistry teacher) describes his interactions with university-based scientists in a positive way that underlines the genuine novelty of the research that his Year eight students are undertaking, and the possibilities for exchanging research ideas as equal collaborators, where university researchers are interested in the students' findings:

We do go to York University to have an aseptic techniques research lab each year for the schools in the research hub…because with our ellagic acid [research], because we knew nothing about it beyond the research done by a Year eight girl, I contact York university's chemistry department and said 'what about this?' and they said, 'no idea at all, but we would be quite interested in your results' and I got the same from Herriot Watt university…so I have got kids doing some research into ellagic acid which two universities are interested it! (Keith, chemistry teacher)

Contact with external partners also supported the positive development of research projects through both drawing in the perspectives of other people and, at the same time, developing the participants' sense of identity as a scientist:

It has been great having other people that can help you and the conferences and all the other bits that go with it have been really good because it kind of highlights the perspectives of other people. (Gordon, physics teacher)

Working with people from IRIS and doing research again helps me feel so inspired…I don't want to lose that aspect of being a researcher, because that is part of my identity as a scientist, you need to be active in research in some way I feel to be a scientist. (Rose, physics teacher)

Teachers and technicians highlighted the value of being able to work with colleagues in other schools for their own professional development and to support the learning of their students, and that these networks were 'vital' in sustaining research projects over a long period of time:

I am working with other teachers in schools, other colleagues at universities, being able to generate those links has been excellent for me and for students, in terms of us being able to go out and experience what it is actually like to actually work at a university, even for a teacher, this is a very powerful experience. (Dean, biology teacher)

I got in contact with a teacher at another school that I 'met' on the webinar and we put some of our students in touch who were interested in the more computer science, programming side of the project, because that's not one

> of my strengths, but it was the strength of the other teacher. (Francis, physics teacher)

Research networks and collaborations were developed through both interactions in-person and virtual connections; for example, many teachers did not have direct, personal contact with an academic in a mentoring role as described by Sally (biology teacher). For these teachers, connections with scientists and other external partners were developed through email, social media platforms and online webinars, as Jane (biology teacher) describes, 'emails with scientists working on the research project have helped me connect with research through people outside of school and this has been positive.' Ellen (general science teacher) also described valuable virtual research networks that she had developed through research projects, 'I have developed contacts through twitter and email, networks who I am in touch, so I feel that I have got a new network and contacts with people that I didn't have before.' The value of webinars as networking opportunities was also highlighted by teachers as a way of demonstrating the scope and value of the research collaboration to students as well as teachers:

> Webinars are really interesting, it gave the students and me a chance to directly ask questions of researchers and other teachers working in the projects, it was so interesting for us to see that there were schools across the UK and beyond, and I think for the students and for me, it clicked that we are part of an international team, a really big collaboration and that is really inspiring, it was a defining moment…we felt as if we were part of something bigger, as opposed to it being just something that we were doing in school as a little bit of fun…setting students up to work together from different locations I think it gives them a realistic idea of what it's like in research as a whole but also it is just very interesting for them and promotes that teamwork that is really important in science. (Francis, physics teacher)

John (physics teacher) also highlighted the value of internet-based networking opportunities for students, supporting the development of online research communities:

> We support students in our school and in other local schools...our projects have lent themselves to developing an online community, where students from five schools working together to build a Minecraft version of the ATLAS detector at CERN...the great thing about it is that it made an online community, so they were from fairly spread out schools, but we didn't really meet together. They just set up a chat room, would talk to each other...the online community really appealed to the students, they were able to keep in touch during the summer holidays. (John, physics teacher)

The place of electronic networks and online communities as a source of professional development for teachers has been recognised in wider research. For example, Hanuscin, Cheng, Rebello, Sinha, and Muslu (2014) and Luehmann (2008) explored the role of blogging as a professional development activity for high school science teachers. Hanuscin et al. (2014) suggest that blogging provides a safe space for science teacher leaders to explore different identities (e.g. as leaders) through sustained reflection and social interaction. Luehmann (2008) also notes the opportunities that blogging brings for both individual reflection and developing communities of practice, where teachers can give and receive feedback and display their competence. Blogging is also a way of building an online community of practice and a support network, where teachers can establish a shared group identity with other teachers that reaches beyond the geographical confines of their classroom and school. Developing shared group membership through electronic networks can be particularly important for teachers seeking to develop their identities where they are perhaps the only subject specialist in their school (e.g. high school physics teacher) or perhaps the only teacher engaged in research projects. In the final section, the ways in which the experience of research projects encourage and enable teachers and technicians to establish and develop their own collaborative networks is considered.

6.5 Establishing and Building Collaborative Networks

Unsurprisingly, teachers and technicians who contributed their experiences had varying levels of confidence in establishing and further developing their own collaborative networks. Some were very confident about establishing networks in order to develop resources and contacts so that they could support school-based research. Keith (chemistry teacher) is a good example of highly confident teacher, able to develop his own networks:

> Simply put, if you don't ask, you don't get. If you start by saying to the commercial world that you are from a school, then they normally want to do something to help... when that first contact has given that lead, you go to the second person and say, 'well this person has suggested that you are the person I might ask for help', then, that establishes a starting point and before you know it the network has begun...it is not being cheeky, it is being confident and just asking for that help and that can often get you to where you want and sometimes to where you are not expecting to be! (Keith, chemistry teacher)

Other teachers shared how they drew on previously established research networks, sometimes developed during their prior experiences as a postgraduate or PhD researcher. For those without a background in postgraduate research but who felt confident to develop external collaborations, some described how research projects enabled their contribution to be valued in their wider school community:

> I think people have taken me a bit more seriously, because I was a first year teacher and they were like, 'who is this girl?', and so working with external partners, external companies it legitimised a lot of what I wanted to do in teacher...I was always comfortable that I could communicate with other people and building networks. (Sophie, biology teacher)

The importance of building research networks was highlighted by Rabail (technician) who described how collaboration through research was a powerful way of demonstrating the diversity of science:

> Research allows you to build networks, to be creative, and this shows that science is not just a solo endeavour undertaken by eccentric white men, it is a collaboration that requires you to develop networks, draw a diverse group of people together. (Rabail, technician)

For some teachers, it was only through the experience of research projects that they felt empowered to develop partnerships and networks that extended beyond their own school community and that these networks helped develop their capacity to teach science:

> Research and working with York University opened our eyes to the fact that we could borrow things from secondary schools, they have more equipment that we would have for class work, but also then we thought, we aren't we asking our local secondary schools about this, especially when we are part of a multi-academy trust, so it has kind of given us the nudge to make links with three other secondary schools in our local area. (Ellen, general science teacher)

> We have had Google hangouts with researchers and children from different schools and they will talk about what they have done, what they have researched, and then the children will ask the researchers and each other questions, it has been wonderful working in that collaborative way, it has been very motivating for the children. (Ellen, general science teacher)

Notably, it appears from these interviews that teachers who are supporting research projects that are outside of their subject specialism (e.g. Francis, Mark and Declan), teachers who are solo members of their departments (e.g. Stephen) as well as general science teachers (e.g. Ellen and Madeleine) highlight the importance of developing collaborations with external partners and experts that benefit both their professional development and the learning of their students. Ellen (general science teacher) outlines the support that these networks can provide for generalists:

> Having other teachers in other schools help support us was so helpful, maybe because we can bounce off ideas, and at the end of this project we got together and said, 'how could we improve it for next time? What else could we do?' So, we're learning together, as teachers, and children are learning together as well. (Ellen, general science teacher)

Teachers who support research projects across a group of schools also highlighted the important role that research networks can have in disseminating subject-specialist expertise to support non-specialist science teachers:

> We have taken research projects into different schools within our MAT and, there are teachers there who aren't necessarily science trained but it has given them another string to their bow, because they are learning science along with their pupils, working with science specialist teachers and university researchers, all in a big network. (Barbara, chemistry teacher)

Francis (physics teacher) also highlights the importance of external support in her role as a physics teacher who does not have a degree in physics:

> I think the networking is hugely important…as a relatively new physics teacher, who is not a physics specialist…it has been so valuable working with other teachers, with IRIS staff and having the opportunity to ask questions. (Francis, physics teacher)

Finally, Stephen (psychology teacher) shares the value of external networks for those teachers who do not have subject-specialist colleagues within their own school context:

> As a psychology teacher, I am in a department of one, so research projects have helped me develop collaborators, helped me to communicate with teachers in other schools and IRIS staff, it was so helpful having that bank of expertise that you can call upon, it has been really helpful…I think there is something important there about collaboration. (Stephen, psychology teacher)

Collaboration is a broad theme that involves teachers, technicians and students working in different and new ways together that contrast with classroom and curricula-based teaching. Teachers and technicians describe their roles in research projects as that of an advisor, coach, facilitator, guide, mentor and team captain and note the different ways that students work together within and beyond their own year groups. A key part of collaboration is working with a range of external partners including scientists, university staff, IRIS staff and teachers and students from other schools. These collaborations can be established in both personal and virtual learning environments. Both types of collaboration are viewed as a positive development for the experiences that teachers and students have of science research and provide the context for the development of skills that are explored in the following two chapters. Teachers and technicians have varying degrees of experience and confidence in establishing and developing their own collaborative networks, and this can be linked to their own previous experience of research and/or building networks. Collaborative experiences provided by research networks are recognised as being particularly valuable for those teachers who are generalists, those who are supporting projects that are beyond their subject specialism and teachers who are sole members of their department. The importance of collaboration is also highlighted in wider research that considers the professional development of science teachers (Mehli & Bungum, 2013; Varelas et al., 2005). As with the teachers and technicians who shared their experiences for this research, other projects such as the Scientist in the Classroom Partnership (SCP) (Ufnar et al., 2012, 2017; Ufnar & Shepherd, 2019) emphasise the importance of providing school communities with access to external, expert STEM partners. Unlike the SCP project where teachers worked with external STEM partners on a weekly basis over the course of an academic year, teachers and technicians working in this study rarely had similar levels of face-to-face engagement with external partners. Furthermore, teachers and technicians involved in this study had a greater range of external partners. Even with these distinct differences in the model of delivery, collaboration is found to be a fundamental and positive aspect of school community engagement with STEM research.

References

Hanuscin, D. L., Cheng, Y. W., Rebello, C., Sinha, S., & Muslu, N. (2014). The affordances of blogging as a practice to support ninth-grade science teachers' identity development as leaders. *Journal of Teacher Education, 65*(3), 207–222.

Luehmann, A. L. (2008). Using blogging in support of teacher professional identity development: A case study. *The Journal of the Learning Sciences, 17*(3), 287–337.

Mehli, H., & Bungum, B. (2013). A space for learning: How teachers benefit from participating in a professional community of space technology. *Research in Science & Technological Education, 31*(1), 31–48.

Ufnar, J. A., Bolger, M., & Shepherd, V. L. (2017). A retrospective study of a scientist in the classroom partnership program. *Journal of Higher Education Outreach and Engagement (TEST), 21*(3), 69–96.

Ufnar, J. A., Kuner, S., & Shepherd, V. L. (2012). Moving beyond GK–12. *CBE—Life Sciences Education, 11*(3), 239–247.

Ufnar, J. A., & Shepherd, V. L. (2019). The Scientist in the Classroom Partnership program: An innovative teacher professional development model. *Professional Development in Education, 45*(4), 642–658.

Varelas, M., House, R., & Wenzel, S. (2005). Beginning teachers immersed into science: Scientist and science teacher identities. *Science Education, 89*(3), 492–516.

7

Professional Development

7.1 Introduction

When reflecting upon their experiences of research projects, teachers and technicians frequently described their work with students as part of their professional development as educators. Keith (chemistry teacher) suggested that research projects had been a 'central part' of his professional development and that he had been able to 'upgrade knowledge and skills'. Likewise, Ellen (general science teacher) emphasised the importance of research projects, suggesting that this experience had, 'changed my professional life' and had 'opened my eyes and put me in touch with other professionals, research projects strip away many barriers and have helped me feel part of a professional body'. Research projects provided some teachers with an opportunity to consider the purpose of education in a broader sense, as Anthony (physics teacher) described: 'research projects have made me think about education in a much wider sense, for example, developing reflective and critical thinking.' Another teacher also reflected upon the experience of research project in the context of wider thinking about the purpose of education and their own professional development:

E. A. C. Rushton, *Science Education and Teacher Professional Development*, Palgrave Studies in Alternative Education, https://doi.org/10.1007/978-3-030-64107-8_7

> My professional development goals have always been geared towards power dynamics, social justice and equity and research projects help me provide greater opportunities for a more diverse group of students in science, opportunities they might not otherwise have. (Jacey, psychology teacher)

In this extract, Jacey links the capacity of research projects to provide opportunities for students to her own professional development as an educator. In the following chapter, the impact of participation in research projects on students is considered in detail. Moreover, broader considerations of research projects as a form of professional development, teachers and technicians identified three distinct aspects of professional development:

1. the development of new or enhanced skills, including practical, interpersonal, and subject knowledge,
2. increased recognition of their role and their subject, including the development of alternative pathways to professional development beyond traditional management routes, and
3. research projects as a pedagogical approach to science education.

As part of their reflections upon the role of research projects in their wider professional development, teachers and technicians also more frequently detailed the challenges they experienced, and these are also outlined in the latter part of this chapter.

7.2 Opportunities to Develop and Enhance Skills and Knowledge

Teachers and technicians described specific skills they had developed as part of their involvement in research. *Practical skills* included using high-quality microscopes, laboratory techniques, and computer-based skills for example, using new software and programming. For example, Sophie (biology teacher) said: 'I have learnt lots of skills, about plants, learning how to do soil testing' whilst Ellen (general science teacher) described how she was 'upskilled' in using microscopes as part of practical work with students. Technicians were also able to expand and enhance their use of practical techniques both in terms of the necessary skills to undertake new procedures in the school laboratory, but also to lead workshops with students so that the students could become proficient in using these practical techniques that were new to them:

> During the research project I have been more involved in setting up practical sessions and ordering reagents, these are skills I would not necessarily have had the opportunity to develop without the research project…my confidence has increased greatly, I now run workshops for biology students as part of their core practical. In the past this was something that the students only did with their classroom teacher. (Bailey, technician)

Software and programming skills were also specifically identified by teachers as skills that they had developed themselves through the course of their participation in research projects. Jane (biology teacher) said, 'the skills that I have gained have been practical, I have learnt to use sophisticated software'; Michelle (physics teacher) stated, 'I have had to develop a load of skills that I didn't have. I have had to get better at programming in order to be able to support my students during the first part of the project.' Research projects also provided teachers with an opportunity to gain 'experience of how research science works' (Jane, biology teacher) and, through this experience, refresh and draw on previously developed research skills. Stephen (psychology teacher) said:

> My skills in terms of researching journal articles and reading them and gathering the relevant information, they were quite rusty, it is a long time since I have studied and researched myself, so research projects have required me to reactivate those skills and reactivate that part of what I can do as a teacher, as a researcher. (Stephen, psychology teacher)

In these descriptions of skills development, there are examples of teachers and technicians describing both the enhancement of previously developed techniques as well as generating skills that are new to them and to their context as educators. Bailey (technician) also described her increased confidence with using practical skills and augmented role with students. In addition to the development of new skills, teachers and technicians identified that research projects had provided an opportunity to develop and improve their subject knowledge. Hasan (physics teacher) stated, 'as a mentor for a variety of projects, I feel this has been a real factor for increasing my own knowledge in so many different fields in science.' Michelle (physics teacher) and Bailey (technician) also described the improvements to their own subject knowledge:

> Ensuring that I was constantly addressing my own subject knowledge, and getting to grips with newer science, that was on the forefront was crucial to me, and it is something that research projects support you to do and this approach to addressing subject knowledge has become part of our department development plan. (Michelle, physics teacher)

> Liaising with university staff and undergraduates has encouraged and motivated me to keep up to date with current science linked with the curriculum. I am able to stretch certain parts of the curriculum to challenge our most able students with confidence. (Bailey, technician)

These reported gains in subject knowledge through participation in research projects is consistent with wider literature that explores gains in science content knowledge as part of teachers' continuing professional development. For example, deeper subject knowledge and greater confidence in teacher understanding of science content were reported impacts of the Scientist in the Classroom Programme (Ufnar & Shepherd, 2019). The importance of subject knowledge development as part of effective teacher professional development has been underlined in numerous studies (Banilower, Heck, & Weiss, 2006; Cohen & Hill, 1998; Garet, Porter, Desimone, Birman, & Yoon, 2001; Kennedy, 1998), and this underlines the value of secure and deep subject knowledge as critical element of school science teaching (Cohen & Hill, 1998; Kennedy, 1998; Ufnar & Shepherd, 2019; Supovitz & Turner, 2000). As has been described above, some teachers and technicians described how their confidence to use, and support student use of, practical science increased as a result of participating in research projects. Increased confidence and other transferable, interpersonal skills identified by teachers and technicians are considered in greater detail.

7.3 Interpersonal Skills

These skills, as described by teachers, included confidence and self-belief, leadership and organisation, reflection and communication (oral and written). Madeleine (general science teacher) described how research projects developed her confidence in her abilities as both a teacher and a scientist:

> My confidence as a teacher has also grown, but maybe more than that, my confidence in my own abilities as a scientist has grown because I have learnt new skills and I have been put back in that place of not knowing and feeling discomfort. (Madeleine, general science teacher)

Sally (biology teacher) also described how the experience of research has increased both her skills and her self-belief, 'research has given me a skills set, an enthusiasm and self-belief that I just didn't realise that I had, maybe I had it all along, but I lost some of it along the way'. Research projects sometimes required teachers and technicians to develop their skills of persuasion, patience and persistence, to encourage others to support their work with students, as Keith (chemistry teacher) highlighted:

> I have got huge skills out of this, skills of persuasion, of getting support and equipment, visits, getting different and changing management teams on board, being patient and sticking with it when it seems like there isn't support…persuading other staff to let me have lab time and technical support. (Keith, chemistry teacher)

Joan (technician) also noted the need to develop logistical and organisation skills to enable students to participate in off-site trips, often at short notice and that this also required negotiating permissions from school leaders. The importance of working within a school context with the support of senior management for research projects is considered in greater detail in the later section of this chapter that explores the challenges teachers and technicians encounter when championing school student research.

Research projects also provided teachers and technicians with a range of opportunities to develop their communication skills that went beyond more frequent school-based experiences. For example, Bailey (technician) described how her work facilitating visits by academics to her school enabled her to improve her ability to facilitate seminars with scientists and students and manage question and answer sessions. Michelle (physics teacher) also had the opportunity to develop her communication skills, she said, 'I had to get more comfortable with speaking to larger and different audiences, even if just to introduce my students when they featured on the local television news.' Involvement in research projects also provided teachers and technicians with opportunities to enhance their written communication skills, authoring articles about their experiences of research and co-devising and co-authoring resources to support other school staff to initiate and implement research projects themselves. Dean (chemistry teacher) suggested that writing about his experience of research was a novel experience:

> In terms of my own professional development it has meant, for example, writing an article for the *Teaching Times* magazine about working with students in research…this was something that I hadn't had the opportunity to do before, and that really pushed me out of my comfort zone, in terms of being able to communicate what I do to a wider audience, which was really beneficial for me. (Dean, chemistry teacher)

Another skill that was less frequently described but was evident in some teachers' experiences of research was the ability to be reflective in their role as a teacher. For example, Rose (physics teacher) suggested that the differences in learning environments that she encountered in the traditional classroom setting compared to working in research projects with the same students (see Chap. 4) prompted her to reflect upon the reasons for those differences and to value the role of reflection in her work as an educator, 'I never used to be this reflective, it is another skill that I have developed. I think it is where teaching should be going…reflection should be seen as a valuable tool set' (Rose, physics teacher).

As has been outlined in the preceding discussion, teachers and technicians sometimes separately described the practical and interpersonal skills they developed through their participation in research projects. There are, however, examples where developing increased confidence and self-belief in an individual's ability as an educator is explicitly linked and underpinned by the development of distinct and new practical skills, such as laboratory techniques or using new computer software:

> My experience with research…the skills I have learnt, using the *Apollo* software, and working with scientists outside of school has helped me understand myself as a professional who is both a teacher and a biologist and this is something that has been really positive for my self-belief. (Jane, biology teacher)

> I think, as a non-physics specialist when I came into teaching, this has had a huge impact on my confidence and also kept my interest in physics alive, and the more I get involved with research projects, the more ambitious I become about what I can offer my students. (Francis, physics teacher)

In addition to skills and subject knowledge development, some teachers and technicians highlighted that their involvement in research projects increased the positive recognition they, and the subject they taught, received from school colleagues and senior leaders, and this aspect of professional development is now considered.

7.4 Increased Recognition of Teachers, Technicians and the Value of Science

The responses of colleagues and managers to teacher and technician involvement with research projects were frequently discussed, with some identifying that their participation in research enhanced their colleagues' perception of their professionalism and abilities because they were creating additional opportunities for their students beyond the curriculum:

> I have noticed a shift with colleagues…other staff in the department, since I have been doing this research, they have seen me do something…that has a broader purpose and I feel that I am seen as more able, my abilities as scientist as well as a teacher have been recognised, I am seen as someone who is willing to develop extra opportunities for students, to go above and beyond, and in a way that is unusual and perceived to be of high academic worth. (Jane, biology teacher)

> The school Special Educational Needs Co-ordinator has come into my sessions to see how the work was going because we have managed to engage some of the students with Special Educational Needs in this project and that is seen by my colleagues as interesting and valuable as some of these students do not usually get involved in this kind of academic enrichment. (Madeleine, general science teacher)

The positive recognition from colleagues is something that Jane and Madeleine had identified during their participation in research, it does not appear that either teacher anticipated or was motivated by this type of external recognition or validation, but that this is a welcome consequence. For others, recognition comes from a formal approach—building in research projects to performance management/appraisal targets—for some this is reactive to alleviate pressure from managers to meet performance management criteria:

> We have a huge amount of pressure with our performance management because I am at the stage which is the highest that you can get to without having any responsibilities you have to fulfil: all these other criteria and lots of them involve whole school stuff or developing students beyond the curriculum and that sort of thing, so this research has been useful for that. (Amy, physics teacher)

For other teachers, this formal approach to recognition was more positively motivated, and this appears to be linked to the level of support that teachers perceive they receive for this aspect of their work with students:

> So one of my appraisal targets is to 'enrich the school experience' so my Head of Department encouraged me to put the research project down as a

way of doing that, and that includes the visitors that come to the school, the trips that the students get and all of that work as well as the project sessions themselves. (Madeleine, general science teacher)

The value of a teacher and their role in schools can be synonymous with the value of the subject that they teach and, for some teachers who principally taught elective subjects (such as psychology and computer science) rather than compulsory post-14 subjects (e.g. mathematics and science subjects), research projects were a way of increasing recognition of the value of the subject they taught and the validation they received for their work with students:

You don't need a psychology teacher in a school like you need a maths teacher, so I have got to work hard to get students to take my subject…there is a problem of the status of the subject and that is a motivation to incorporate research projects, to raise the profile and value of my subject, it is something that I can offer from my subject to students across the school. (Stephen, psychology teacher)

Working on the project has been great, what has been really helpful is getting some validation for the work I am doing, sharing that learning with other teachers through webinars, the validation from other teachers saying that the resources are helpful, that is huge because particularly in my subject in a school context you are often working down the salt mines by yourself without colleagues or a team, so research projects help in that. (Jacey, psychology teacher)

Teachers and technicians also described how they explicitly used research projects to raise the profile of the subject they taught within their school contexts and to encourage more students to study science post-16 years:

The involvement of CERN has been massive on the popularity of physics and to be able to carry out a research project in school which links with the scientists in Switzerland has raised the profile even further. (Joan, technician)

> Part of our school improvement journey is about raising the profile of subjects such as science and developing research projects has already led to the profile of STEM being raised in our school…this is crucial for us in terms of trying to attract students to the school and also to raise the profile of STEM in the local community. (Dean, biology teacher)

Some teachers who had been involved in research activities for more than two years identified leading research activities as a professional development opportunity that was an alternative to progression through management pathways such as Head of Department, or Deputy Headteacher roles:

> I haven't made any progression into management…I don't want to be a manager, but I do want more responsibility and to run something at a whole school level and research across a school could be a mechanism towards that, giving me time and resources to do that. (Sally, biology teacher)

Furthermore, some teachers who have had experience of school management roles and leading research projects suggest that they have actively chosen to continue their school-based research with students rather than pursue leadership roles, further highlighting the value of research projects providing an alternative pathway for teacher professional development:

> For me, I did management, I was on the Head of Science track, then Deputy Head track, but when I got funding to spend one day per week supporting research projects, I had to make a choice. This way of teaching is a different approach to progression and an alternative to management. (Keith, biology teacher)

> I have actively decided not to pursue a career in management as I wouldn't have time to do research projects as well as that role…a couple of years ago our Head of Department was ill and I stepped in as acting Head of Department, and so I know I would not be able to combine the two roles, and I didn't go into teaching to do a management role. (Mabel, biology teacher)

Sally (biology teacher) suggests that having a family has been a barrier to her career progression and recognises that the full-time nature of management roles has prevented other women from progressing into leadership roles in schools. Sally contends that research is a way of using the capacity and capabilities of 'really clever women', as research can be done in a part-time role and more flexibly than management and this would increase the quality and capacity of the teaching workforce:

> The pathway to management and leadership doesn't always fit very well with families and there is all this unused capacity in the teaching workforce that we are just not using, the research takes a lot of time, but you can do it part-time and management roles are very difficult to do part-time, but there is no reason why you can't do research in a part-time role, because, you do feel left behind, I mean, I have got my friends from school in other professional roles who have had children and done part-time work but they are still creeping up the ladder because they are able to progress through their career whilst still working part-time whereas I don't feel that I could… I think in teaching you have to make that call, that choice. (Sally, biology teacher)

Clearly, for Sally, Keith and Mabel, research projects have provided an extended framework of professional development that they equate to roles including Head of Department and Deputy Head. Whilst this is not a universal experience for the teachers and technicians who shared their experiences as part of this research, many do identify that research projects enhance their professional development through the generation and augmentation of a range of practical and interpersonal skills and it is possible that over time this could be extended in to more extensive professional development. As well as providing opportunities for increased recognition for the value of the subject they teach, research projects provide a framework for further recognition by colleagues and school leaders of teachers' individual contribution either informally or through documented appraisal processes. As has been explored in Chap. 4, teachers frequently described the variety of teaching and learning methods and approaches that they could incorporate into the implementation of research projects, and that this way of working with students provided 'freedom' in comparison with curriculum and classroom contexts. When discussing their professional development, teachers described how research projects provided a context for developing their pedagogy, both in general terms and especially research projects as a pedagogical approach to teaching science in schools, and it is to this aspect of professional development that is now considered.

7.5 Developing Teachers' Pedagogical Understanding and Approaches

Teachers regularly suggested that the opportunity to work in research with their students had 'refreshed' their teaching and that it was important to regularly take up opportunities to do this otherwise they could become 'stale':

> If you want to remain a good teacher you have to keep refreshing what you are doing, trying new things, reflecting, because otherwise you get stale, and if you get stale you get bored and you become boring and your lessons are very flat…I did a PhD…but my involvement was nothing to do with my former research background, my involvement is about exciting young people about biology and to do that you have to be excited yourself. (Peter, biology teacher)

Peter argues that refreshing teaching is necessary to be considered a 'good' teacher, and that refreshment should bring excitement to both the teacher and the student when learning about science. Marina (general science teacher) suggests that this idea of 'refreshment, excitement' could cascade into other teachers, only involved on the periphery:

> These projects, whether you are leading them, or involved as a supporter, these are what make teaching so exciting, not teaching the same classes the same things, I mean you try and keep it fresh and interesting but it is hard, marking books is not exciting, but this, the project work is really exciting. (Marina, general science teacher)

Gordon (physics teacher) also links his professional enthusiasm and excitement as a teacher with student engagement: '[research] keeps me enthusiastic, and of course that is going to spill off onto the kids.' As has been discussed in Chap. 6, experience of research projects has led other teachers and technicians to reflect upon the different roles they have as a teacher, and how understanding these roles helped them to better respond to the needs of their students. Both Keith (biology teacher) and Clare (chemistry teacher) suggest that this augmented understanding of the way students work extends their professional development, and encourages them to give students more freedom and responsibility to discover answers for themselves:

> The way I think is completely different to when I am in class compared to research projects, projects allow students the freedom to show us what they think, or how you think, once it is handed over to them and they get their legs and they are confident to pursue, it is breath taking and I know that they would quite happily skip lessons to do this all the time. (Keith, biology teacher)

> I guess the research makes you think about your role as a teacher a little bit more, it has made me more aware of the different role that I take, and I consciously take a decision not to tell the students the answer, even if I have gone away and found it out, they are not getting it out of me, it is for them to discover for themselves. (Clare, chemistry teacher)

As well as the general areas of 'refreshing' teaching, enthusing teachers in their role as educators and providing the stimulus to reflect upon the differences in practice in and beyond the classroom, teachers described in detail how their experiences supported the development of school-based research projects as a pedagogical approach to teaching and learning school science.

7.6 Research Projects as a Pedagogical Approach to Learning Science

Unlike similar studies, where school science teachers expand their understanding of inquiry through collaborations with university-based scientists and researchers (Ufnar & Shepherd, 2019), teachers involved in this current study did not receive any explicit pedagogical content knowledge enhancement as part of their engagement with research projects. Despite this difference, teachers who contributed their experiences as part of this research frequently identified how research projects had developed their approach to teaching science in the classroom and specifically how this refocused science teaching on practical science and inquiry:

> Rather than teaching with a focus on the objectives that I have got cover with the students, I am bringing in a more over-arching sense of scientific enquiry, and this is more purposeful and the scientific enquiry skills are more embedded. (Ellen, general science teacher)

> Research projects have really changed the way I teach, because I can now, when I'm teaching other parts of the course, I can say, "Look, this is what we did with our project. Look at our results, and this is how we analysed it". That has been really helpful for my teaching, and for them, I think, as learners because we are focused on practical science, on inquiry. (Sally, biology teacher)

Sally (biology teacher) suggests that she actively transferred the pedagogical approaches she identified as being successful in research projects to the classroom context and relatedly Hasan (physics teacher) describes research projects as providing a framework for teachers to become 'gradually brave enough to move toward unknown borders of science teaching…moving away from 'safe and secure' prescriptions in their science teaching practice.' Similarly, Sophie (biology teacher) described how research projects shaped her teaching, particularly developing her skills in questioning, and her experiences align with Hasan's suggestion that research projects enable teachers to develop into new areas. Nathan (biology teacher) described how students themselves transferred their approach to science developed in research projects to learning science in the context of the classroom:

> Having authentic science research projects and regular scientific discussions is fostering a love for learning…the class discourse is changing from the default exam focused questions to more inquisitive questioning which enriches their understanding of the subject as a whole. (Nathan, biology teacher)

For some teachers, incorporating experiences of research projects into classroom teaching was a process that occurred over time:

> In the first year of running research projects I saw them as very separate from what I was teaching in the classroom, but I think with the project this year I have been trying to link it more to what we are doing in lessons…for example when I was teaching the life cycle of starts I felt more confident as I had more real life examples and I could use actual data, in terms of the data that telescopes get. (Francis, physics teacher)

Here, Francis, Sally and Nathan describe how they incorporate approaches developed as part of research projects to their classroom teaching of physics and biology, and Nathan also highlights how students themselves approach their classroom, curricula lessons with a greater focus on inquiry questions, rather than identifying answers to exam questions. These two approaches to learning, inquiry focus *versus* exam focus, are also found in Keith's description of his students' experiences; however, unlike Nathan, Keith suggests that students move between these two approaches depending on the context and that the inquiry approach found in research projects does not move into the curricula classroom context:

> You can see the gear change is fantastic from research projects to lessons, the students will be doing their research at lunch time, they will be thinking and planning out their experiments, then the bell goes at the end of lunch and you get the gear change, put all the stuff away, get the textbook out and they immediately start copying out the key terms and answering the summary questions and they switch into school, exam focused think-

ing as opposed to pilot studies and having fun with research. (Keith, biology teacher)

Some teachers took their understanding of how students learn during research and developed their teaching, most notably in the teaching of practical sessions and in more conceptually demanding curriculum content, such as radioactivity and genetics. Natalie (chemistry teacher) suggested that it was only during her experience of research with students that she could see the struggles students had with experiments and stated that this challenged her to better understand student engagement in classroom practical sessions:

> Having seen how students have struggled with initial sort of practical troubleshooting in the research group time, techniques that previously I through were simple and easy, but are really not tackled well at school, I have adapted certain things at A-level, more than anything it has really made me interrogate how I deliver practicals. (Natalie, chemistry teacher)

Elliott (physics teacher) suggested that his involvement in research projects changed his understanding of what was possible in terms of students' participation in research and the ways in which students can develop their knowledge and understanding of scientific inquiry, 'to discover deep scientific principles through seemingly simple practical tasks, as part of students conducting their own personalised projects.' For Gordon, a physics teacher of over 30 years, using a detector that displayed alpha, beta and gamma radiation particles on a computer screen had a significant impact in both the way he taught radioactivity and the ability of students to grasp an abstract concept in a 'visual' and 'real' (i.e. apparent to the senses) way that made it easier for them to understand the concept and progress to posing questions about radioactivity:

> For radio activity there are very few experiments you can do…but if you have got something on your desk and it is showing radio activity with pictures of particles whizzing on the screen…somehow it becomes more comprehensible to the students because…it is shown as an image they can see in real time…and therefore it is much more real to them…so the detector

> has transformed the way I teach and the way the students can conceptualise and understand radiation…this detector allows you to ask questions—'what happens if we go outside? What happens if we go to a different classroom?' The previous kit was just a demo, this kit enables the students to ask questions about radiation and that surely means that they understand it better. (Gordon, physics teacher)

Making an abstract concept visible and therefore more comprehensible was also a feature of the development of biology A-level teaching for teachers and technicians involved in the *Genome Decoders* research project. As with teaching radioactivity, incorporating research into the teaching of genetics enabled teachers to develop their pedagogy so that students were motivated to develop both knowledge and skills, and bring them together to develop a deeper understanding necessary to progress in the research project:

> The project has been hugely important for how I teach genetics, which is part of the A-level biology course, as it gives the students the chance to learn abstract concepts in practical ways …they were able to connect aspects of their knowledge to the skills they had developed, and they were so enthused and motivated to keep going. (Jane, biology teacher)

> This project is a fantastic tool for teaching, taking that very abstract idea [DNA], the students take it on board [in lessons] but it is not gripping in the way that the research is. (Peter, biology teacher)

For Jane, the practical nature of research supported her students' understanding whereas for Peter, the 'gripping' nature of research seemed to drive the engagement of his students, motivating them to better understand DNA. Gordon (physics teacher) suggested that research projects provided him as a teacher with more information regarding the attitudes of students towards physics, particularly when the subject becomes optional, post-16 years:

> These projects gave me an insight into what motivates students to pick physics. The high achievers want to delve into the world of sub-atomic physics and have chosen physics as their University first choice subject. The

younger students were split by gender on why they study physics with the medical uses being particularly attractive to girls. I will be using this as a selling point in the future when it comes to subject choices. (Gordon, physics teacher)

Although from different perspectives, Keith (biology teacher) and Stephen (psychology teacher) strongly stated the value that research projects have as a pedagogical approach to science education. Keith highlights the value of this approach for both students and teachers:

> I want to get this kind of approach to learning recognised and valued by the Department for Education, not just a tick box but that this approach is valuable and develops real skills. We don't value skills in this country, and we should because we really need them…I want the government to recognise the enrichment these kinds of activities give to teachers and students. (Keith, biology teacher)

Simon emphasised that this approach to learning is more akin to that of an academic discipline as opposed to a school subject, and that this integration of knowledge, research methods and practical skills has value for students and teachers:

> The research project has helped me show students how to study as part of an academic discipline, you have to have the body of knowledge, skills, research tools, you need a whole suite of things to properly study something as a psychologist or biologist rather than learning the A-level course…I am trying to bring the university style teaching into my school teaching because I think it leads to better outcomes for students…it has a huge impact because I think it make the topics much easier to understand…this is information you can't get from a text book. (Stephen, psychology teacher)

This description of research projects as an approach to learning science in school is distinctive in that Stephen describes how he is seeking to bring 'university-style' teaching to his approach because he argues that this approach provides better outcomes for students. Previous research has explored in detail the range of positive outcomes for students who participate in research projects as part of their experience of science education (Bennett, Dunlop, Knox, Reiss, & Torrance Jenkins, 2018). These include increased student attainment (Burgin, Sadler, & Koroly, 2012; Daly & Pinot de Moira, 2010; Krajcik & Blumenfeld, 2006; Rivera Maulucci, Brown, Grey, & Sullivan, 2014; Sahin, 2013), improved student attitudes towards science (Moote, Williams, & Sproule, 2013; Welch, 2010; Yasar & Baker, 2003) and increased aspirations towards careers in science (Adams et al., 2009; Hubber, Darby, & Tytler, 2010). This current study makes an important contribution to understandings of the impacts of research projects in that it provides detailed insights as to the experiences of teachers and technicians who participate and, as has been explored in this chapter, the role of research projects in their professional development. Although the vast majority of teachers and technicians described research projects as a positive part of their role as educators, they also outlined a range of challenges and barriers that they encountered as part of their experiences. In exploring the nature and scope of these challenges across a variety of school contexts and research projects, it is possible to provide some further insights as to the ways in which both individual educators and schools as a whole can be better supported when they engage in research projects with students and external partners.

7.7 Understanding the Challenges of School-Based Research Projects

In the following discussion, the circumstances and contexts that support teachers developing research projects, from the periphery of their practice as educators to that which comprises an integral part of their professional development, are explored. Why are some teachers and technicians able to incorporate research projects as part of a pedagogical approach to science education that extends across curricular and extra-curricular teaching and learning, whereas for some these activities are contained within perhaps weekly lunchtime sessions over the course of a single school term? The simple answer is that some teachers encounter more barriers and challenges whilst other teachers work within more supportive contexts. Challenges include:

1. teacher workload and time constraints;
2. teacher knowledge and skill;
3. logistics; and
4. support provided by school leadership and senior management.

In the following discussion, these challenges, and others, are considered and, in the final part of this chapter, some key areas of practice will be presented to provide teachers with support and guidance when incorporating research projects across their practice as educators who mentor school student research.

7.7.1 Teacher Workload and Time Constraints

Teachers frequently commented that their workload was a significant challenge for example, Stephen (psychology teacher) said, 'I am overwhelmed by workload, I can't do everything, all of the time.' Workload was described as challenging both in terms of finding the time to deliver research projects as part of their extra-curricular offer but, also having the time to develop and implement the projects in a way that was 'comprehensive' and 'properly done', particularly in the early phases of new projects, when challenges would inevitably arise and with greater frequency compared to when projects were well-established. Mark (physics teacher) described these challenges that are especially associated when establishing a research project:

> The biggest challenge or barrier is staffing, we had a teacher who was going to lead on one project and at the same time they were given a role as a Head of Year and that was too much of a workload, and the teacher really needs to get the project kicked started, and then if you have a busy teacher, who doesn't have a lot of time and they hit a bump in the road, for example an issue with software, rather than solve this issue, the teacher will often back away and then the project does not get off the ground. (Mark, physics teacher)

Michelle (physics teacher) highlighted that once the teacher had supported research projects through the initial phase, the time commitment from the teacher reduced significantly and this matches Mark's perspective that staff time is needed at the outset to get the project 'kick started':

> What it takes at the start is that investment of time, your time, students' time, building their resilience, to a certain extent, but after that I would say that it is not that time consuming, because students quickly become independent researchers. (Michelle, physics teacher)

Some teachers recognised that they often worked on projects beyond their contracted hours and for some this was only possible because they had part-time teaching roles, so were able to use their 'free' time to deliver projects, for example Bailey (technician) described how in busy periods she would come in on her day off in order to support students, particularly if they were preparing for an external event such as a conference or a workshop with a primary school. Other teachers, particularly those who had experience of leadership roles within schools, suggested that if staff saw the value of projects for their students, they were willing to give of their time:

> Time is available if the project is right: We're seeing teachers giving up their time, and this is teachers working in a profession who have a huge workload, so for teachers to willingly give more time, that just shows how powerful this sort of project is to teachers. (Barbara, chemistry teacher)

Of the 53 teachers and technicians who contributed their experiences for this study, 8 had been allocated time by school management to deliver research projects with students within and beyond their schools. This time was sometimes provided through a temporary reduction in teaching load, for example, 'releasing' a teacher for a half day or day a week over the course of a term or an academic year so that they can establish research projects. Other teachers had their planning and marking workload reduced over similar periods of time to provide them with enough capacity necessary to develop and deliver research projects. As Anthony (physics teacher) described, a temporary reduction in teaching load was critical in enabling him to develop the networks necessary to establish and maintain research projects:

> I've been given a day a week off timetable for a year, which lets me set up and develop projects with students and build networks with my local schools and universities. That time has been a great thing for me, before that it was quite an uphill struggle because you're always trying to slot things in where you can, do things in your own time…now I have got a bit more freedom and I am really grateful for that, but it is with the support of my Headteacher and Head of Faculty because they see it as a valuable thing. (Anthony, physics teacher)

In addition to workload, teachers also highlighted the time pressures associated with incorporating research projects alongside the demands of a 'content heavy' curriculum and within the 'constraints' of external examinations. However, some teachers argued that it was possible to incorporate research projects as part of the curriculum and that this meant they were able to reach more students than was possible if research projects were solely an extra-curricular offer. Associated with the challenge of teacher workload is the time teachers and technicians require to develop the knowledge, skills and expertise necessary to support student participation in research, with many teachers highlighting how difficult it was to find sufficient time to develop new skills including programming and using new software. Madeleine (general science teacher) described how having the time available to work with the first cohort of students was especially important as both she and the students were learning new skills and ways of working, for example, using unfamiliar microscopes and that this experience inevitably took more time than in future phases of the project. Bailey (technician) outlined how, at times research projects required her to support students with techniques and concepts that she herself was unfamiliar with, and this has proved hugely challenging and at times meant that the project had become 'too ambitious' and that she required greater support from the project's university partners. As well as the challenges associated with time and workload, teachers and technicians detailed the logistical barriers they often encountered.

7.7.2 Logistical Challenges

The unreliability of school IT systems posed an often-insurmountable barrier to participation in projects that required students to access data via software packages, and some teachers shared their deep frustration that young people had not been able to engage in certain projects due to these technical challenges. Relatedly, teachers reported that they frequently had to find solutions to unpredictable technical issues so that they and their students could participate in online webinars that they recognised were a valuable part of networking with other schools and students engaged in research projects. Timetables and the inflexibility of the school day also presented persistent challenges for teachers and technicians in terms of finding sufficient and regular time to devote to research projects with students as a group and also having computer room or laboratory space to house sessions. Keith (biology teacher) described how changes to the duration of lunchtime placed a significant constraint on the time students had available to undertake research projects and that it was difficult to find an alternative time where both students and staff were available:

> The biggest barrier…was the shortening of the lunchtime, this was done to improve behaviour, and so our lunchtime clubs, which ran very successfully, they were squashed into a very small space of time, 50 mins to get to lunch, eat and then get to research club, so we really on have 20 mins at the most, which is no time at all…I tried to increase the time available for research by holding after school sessions, but the pressures on the curriculum and on results means that the science department offers interventions on nearly every night after school, or I have a meeting, so there isn't a regular weekly slot where we can do work with the students. (Keith, biology teacher)

Coupled to this challenge of identifying a regular meeting time after the school day had finished was that some students, particularly those in rural areas where transport is limited, were unable to participate in activities beyond the school day because they were reliant on the single bus that departed at a fixed time. Mabel (biology teacher) suggested that a way to overcome this barrier would be the provision of flexible afterschool transport that could be booked a few days in advance so that students could incorporate afterschool research project sessions into their weekly timetable. Peter (biology teacher) highlighted how the timing of projects over the course of an academic year was also important so that these activities were 'balanced' alongside the demands of public examinations and university applications. Ralph (physics teacher) also described how the changing nature of post-16 education in schools provided greater logistical challenges, principally that students have less contact time and different expectations of the skills and experiences they should gain during this phase of their education which makes it difficult to incorporate research projects as part of the timetable:

> One of the biggest challenges is that post-16 education is changing, it is much lighter on contact time, it can be quite part-time, and some of our students have 16 hour per week jobs, so there is much less opportunity to have them in school working on research projects, and there is the idea that being economically active earlier on is a good thing, gives good life experience, so a very different ethos from full-time education where you get here at 8.30 am and stay until 3.30 pm unless there is a specific reason…it is much harder to integrate sixth form extra-curricular. Also, sixth form education is a loss leader, yet compulsory, that combination of factors can make it different and challenging. (Ralph, physics teacher)

The lack of opportunity for regular contact time is especially pertinent in light of previous comments from teachers about the importance of having sufficient time to persevere through the particularly challenging early phases of the research project. Anthony (physics teacher) also described the challenges he encountered in supporting students' sustained engagement once the initial interest and enthusiasm has faded and how his role as a teacher was to encourage students to 'keep going' and 'understand that the nature of research is about being consistent and plodding away until you come up with something that is useful and meaningful'. Teachers and technicians far less frequently described engagement with external partners as a barrier, indeed, as has been documented in Chap. 6, opportunities to collaborate with partners based in universities and research institutes was almost universally identified as valuable. One biology teacher, Sophie, suggested that working with a partner such as an outside company was more challenging as, 'the timelines for a government school and a private company didn't always match up' and that co-ordinating the expectations of schools and companies required 'teacher management'. A key source of support or conversely challenge that was discussed by many teachers and technicians was that of school leaders and managers, and this is now explored in more detail.

7.7.3 School Senior Leadership Support for Research Projects

Teachers and technicians described how the support of their immediate line managers (e.g. Head of Department) and school leadership team (e.g. Headteacher, Board of Governors) was critical in overcoming the challenges of workload and logistical barriers previously described. Sophie (biology teacher) said:

> Crucially for me I had the Governors, the Principal of the school on board, and they gave lots of help because they understood the value of the work and were willing to support it. (Sophie, biology teacher)

Joshua (biology teacher) also described how the 'support and encouragement' of both a teacher's Head of Department and colleagues was 'really important' to get a research project established and successful and Madeleine (general science teacher) agreed:

> One of the most important ways to help teachers to do this is to do it in the context of support, so my Subject Leader is very supportive, she gave me the time to do this project so I have time to plan properly, it is so important to be part of an effective and supportive team, the other teachers don't necessarily have to do the project but they need to see the value in it and support you, especially in the early phases when these things go wrong, you have to have that environment where you can try things and it not go perfectly and then not be constantly worried that you are going to be judged for it. (Madeleine, general science teacher)

Ned (chemistry teacher) clearly outlined the specific support that a teacher needs in the early phases of establishing a research project, but that this is often hard to achieve in the context of the demands placed upon teachers, students and the wider school system:

> It was so hard at the beginning, but we got through it and that was when I needed the support from my Subject Leader, their belief that the research project was a good idea and that we would learn something…but in the pressured contexts of schools, where there is so little time and so much content to get through there is not often the support or the capacity for a teacher to get through that pain barrier and have that experience and I think that diminishes the experience for the teacher. (Ned, chemistry teacher)

Sarah (technician) stressed the importance of school leadership seeing the 'value' of research projects if she is to gain support in terms of time and resources for additional opportunities related to research projects such as trips and visits by external speakers. Keith (biology teacher) reported that although his school management had initially allowed him the time to run research projects with his students, he describes how changes to the school timetable are creating a practical barrier to carrying out research:

> Will this run next year? I don't know, I am struggling to see how it will fit…because of the restructuring of the new timetable and everything has been tightened up and all the demands made by management and I don't see how, at this moment in time that it can continue. (Keith, biology teacher)

Keith also describes his school management's negative response to his proposal that younger students involved in research projects complete an Extended Project Qualifications (EPQs). The barriers facing Keith (biology teacher) in delivering research with students are philosophical as well as practical. Keith describes a concern that students aged 14 years were not able to cope with qualifications aimed at students aged 17 and 18 years, even though as their teacher, Keith believed they have the skills and capacity to complete them, and furthermore, senior management contended that if students were able to complete these qualifications earlier, they would lack purpose in their post-16 studies:

> I wanted to get my Year 9s to do a EPQ because to be honest they have the material and the experience but management didn't like that, couldn't see beyond it being an A-level qualification and was that too much pressure on them, and what would they do in sixth form? (Keith, biology teacher)

Keith also identifies that he has a different approach to teaching and he attributes the questioning aspect of his teaching to his identity as a scientist and a researcher and that this aspect can be seen by other teaching colleagues as a criticism of their approach and methods:

> I guess I am a bit difficult, I try to be engaging and approachable about it you know, try and get people to see my way of thinking but it is hard. Teachers can be a bit rigid in their thinking and their approaches, asking questions can be seen as a criticism and I don't understand that. Maybe that is the researcher in me, the scientist in me, I am constantly curious and that gets me into trouble sometimes, well no, not trouble, but it can be a bit tricky! (Keith, biology teacher)

Natalie (chemistry teacher) also identified tensions that can occur between teaching colleagues when some are involved in research, with those teachers who are not research active feeling inferior and worried about how they will be perceived in comparison to colleagues with research experience. Natalie suggested that for teachers to successfully integrate as part of the science community, there needs to be a positive, collaborative approach where teachers' perspectives are valued:

> I think one of the dangers of teachers being part of the science community is that…there are barriers to break down there, and [developing an] understanding that this is a process where we all want to work together, and we are not trying to reveal the imposter. (Natalie, chemistry teacher)

James (physics teacher) described research work with students as 'quite isolating', especially when there are few colleagues who are 'interested or willing' to explore this approach themselves. Joan (technician) also stressed that research projects were not something that everybody could or should participate in, but that she had 'tried repeatedly' to persuade her colleagues of the benefits to both themselves and their students. It is this idea of benefit or contribution that research projects make to students, as well as the contribution that teachers and students can themselves make through research to both science and wider society, that is the focus of the next chapter.

References

Adams, E., Ward, T. J., Vanek, D., Marra, N., Hester, C., Knuth, R., et al. (2009). The Big Sky inside. *Science Teacher, 76*(4), 40–45.

Banilower, E. R., Heck, D. J., & Weiss, I. R. (2006). Can professional development make the vision of the standards a reality? The impact of the national science foundation's local systemic change through teacher enhancement initiative. *Journal of Research in Science Teaching, 44*(3), 375–395.

Bennett, J., Dunlop, L., Knox, K. J., Reiss, M. J., & Torrance Jenkins, R. (2018). Practical independent research projects in science: A synthesis and evaluation of the evidence of impact on high school students. *International Journal of Science Education, 40*(14), 1755–1773.

Burgin, S. R., Sadler, T. D., & Koroly, M. J. (2012). High school student participation in scientific research apprenticeships: Variation in and relationships among student experiences and outcomes. *Research in Science Education, 42*(3), 439–467.

Cohen, D. K., & Hill, H. C. (1998). *Instructional policy and classroom performance: The mathematics reform in California (RR-39).* Philadelphia: Consortium for Policy Research in Education.

Daly, A. L., & Pinot de Moira, A. (2010). Students' approaches to learning and their performance in the extended project pilot. *The Curriculum Journal, 21*(2), 179–200.

Garet, M. S., Porter, A. C., Desimone, L., Birman, B. F., & Yoon, K. S. (2001). What makes professional development effective? Results from a national sample of teachers. *American Education Research Journal, 38*, 915–945.

Hubber, P., Darby, L., & Tytler, R. (2010). Student outcomes from engaging in open science investigations. *Teaching Science, 56*(4), 8–12.

Kennedy, M. M. (1998). *Form and substance in in-service teacher education* (Research Monograph No. 13). Arlington, VA: National Science Foundation.

Krajcik, J. S., & Blumenfeld, P. (2006). Project-based learning. In R. K. Sawyer (Ed.), *The Cambridge handbook of the learning sciences* (pp. 317–333). Cambridge: Cambridge University Press.

Moote, J. K., Williams, J. M., & Sproule, J. (2013). When students take control: Investigating the impact of the CREST inquiry-based learning program on self-regulated processes and related motivations in young science students. *Journal of Cognitive Education and Psychology, 12*(2), 178–196.

Rivera Maulucci, M. S., Brown, B. A., Grey, S. T., & Sullivan, S. (2014). Urban middle school students' reflections on authentic science inquiry. *Journal of Research in Science Teaching, 51*(9), 1119–1149.

Sahin, A. (2013). STEM clubs and science fair competitions: Effects on post-secondary matriculation. *Journal of STEM Education: Innovations and Research, 14*(1), 5–11.

Supovitz, J. A., & Turner, H. M. (2000). The effects of professional development on science teaching practices and classroom culture. *Journal of Research in Science Teaching, 37*(9), 963–980.

Ufnar, J. A., & Shepherd, V. L. (2019). The Scientist in the Classroom Partnership program: An innovative teacher professional development model. *Professional Development in Education, 45*(4), 642–658.

Welch, A. G. (2010). Using the TOSRA to assess high school students' attitudes toward science after competing in the FIRST robotics competition: An exploratory study. *EURASIA Journal of Mathematics, Science & Technology Education, 6*(3), 187–197.

Yasar, S., & Baker, D. (2003). *Impact of involvement in a science fair on seventh grade students*. Paper presented at the Annual Meeting of the National Association for Research in Science Teaching, Philadelphia, PA, USA.

8

Student and Societal Development Through Research

8.1 Introduction

In their descriptions of, and reflections upon, experiences of research projects, teachers and technicians frequently attributed their motivation and initial impetus to participate to the students they teach. Peter (biology teacher) and Sally (biology teacher) linked participation in research as a way of keeping their teaching fresh, Keith (chemistry teacher) described how research protected the time he has with students from other aspects of his job and Dean (chemistry teacher) suggested that it would be 'remiss' of him as a teacher not to respond to the interest of his students in research. Sarah (technician) described students' 'enthusiasm and passion' for research and emphasised that it was the students' energy and appreciation for all the opportunities that Sarah was able to provide through research projects that ensured that she persisted through periods of challenge. Teachers and technicians saw research projects as a way of developing their students' inquiry and communication skills and that research projects provided experiences where students could make wider connections, establish networks and gain experience of science research and careers. Beyond individual student development, research projects

E. A. C. Rushton, *Science Education and Teacher Professional Development*, Palgrave Studies in Alternative Education, https://doi.org/10.1007/978-3-030-64107-8_8

provided a pathway for young people and their teachers and technicians to provide solutions to problems that impact the wider world, such as disease, biodiversity loss and climate change. Before considering these broader societal challenges, the opportunities for student development are explored, beginning with the development of inquiry skills.

8.2 Development of Students' Inquiry Skills

The opportunity to develop practical skills in a laboratory setting was frequently described by teachers and technicians as an important way of developing a wider skill set around problem solving and scientific inquiry and that developing proficient use of equipment such as a microscope supported students to think more scientifically:

> The students who were part of the research project used the microscope far more than those students who had just used them as part of curriculum lessons, and of course the research project students had developed more skills but it was more than that, the students understood that the microscope was a tool that could be used to find out answers, it was a part of trying to find out a bit more and that it might not go perfectly the first time, so they need to take time, to be careful and patient and to keep preserving, so it is that practical skill but also more of an intellectual jump that you understand how to use to the tool as part of scientific inquiry. (Madeleine, general science teacher)

The development of scientific questioning and thinking through extended experience of practical skills was apparent across the high school age range. For example, Ellen (general science teacher) described how younger students (aged 11–13 years) developed skills including observation and classification through their experience of research projects:

> They learnt about scientific skills such as measuring, cataloguing, sorting, classifying and observation skills, using new equipment like microscopes and magnifying glasses and then they used new techniques like disaggregation, and they were formulating hypotheses, asking questions. (Ellen, general science teacher)

Similarly, when working with students aged 16–18 years, Nathan described how students' experience of practical research provided the context for scientific discussion and supported students' understanding of challenging knowledge that was novel for students:

> What is particularly noteworthy is the ability of the students to understand, new to them, complex biological processes independently from reading academic journals, understanding very complex science which before the project I would not have thought possible. Having authentic science research projects promotes regular scientific discussions and inquisitive questioning because the practical work provides the scaffold for the conceptual science to be made visible to the students. (Nathan, biology teacher)

Here, Nathan and Ellen suggest that research projects provide opportunities for students to undertake practical work and that through this, students develop broader inquiry skills. Clare (chemistry teacher) identifies that students develop their evaluation skills through research and are more able to understand the quality of information they are presented with:

> I think through this research experience students are more aware of the quality of answers…I think this research project has got them thinking about the quality of the science a little bit more. (Clare, chemistry teacher)

Likewise, Mabel (biology teacher) highlighted that research projects supported teaching around the role of evidence with students across the school, 'I used experiences from the research project to share in a school assembly about evidence, the importance of making decisions based upon evidence', and Rabail (technician) emphasised that research projects enabled 'young people to be scientifically literate and questioning citizens.' Elliott (physics teacher) suggested that research projects enabled students to 'begin to think like a scientist and develop problem solving skills and abilities that employers desire.' In their systematic review of the literature that considers Independent Research Projects (IRPs), Bennett, Dunlop, Knox, Reiss, and Torrance Jenkins (2018) found that these are beneficial for students in terms of the learning of science ideas and the development of a range of skills, including practical skills, and these benefits appear to be replicated in the experiences shared by teachers for this current research. This is also consistent with the findings of Dunlop, Knox, Bennett, Reiss, and Torrance Jenkins (2019), which suggests that teachers view open-ended investigative project work which has aspects of student autonomy as highly beneficial for students' skill development and understanding of inquiry. In addition to the specific skills of scientific inquiry and practical and investigative work, teachers in this study highlighted the ways in which research projects extended their students' communication skills, and these are now considered.

8.3 Development of Students' Communication Skills

As has been outlined in Chap. 3, there is a diversity of research projects that provide the context for teachers' experiences, and projects vary in terms of the scientific foci, involvement of external partners, the extent to which they are student-led, the length of project and the age of the students who participate. Opportunities for students to communicate their work also vary; however, there are three formats that teachers commonly describe:

1. oral presentations, for example at a student conference;
2. poster presentations shared at a student conference and then displayed in school; and
3. community events which include workshops with younger students and presentations to parents or school community groups.

Less frequently described are opportunities for students to author and co-author research papers for publication in peer-reviewed journals; however, there are notable examples of this, especially linked to physics projects (Hatfield, 2010; Whyntie & Harrison, 2014, 2015) and evaluating the impact of research projects on students (Parker, Fox, & Rushton, 2018; Rushton, Charters, & Reiss, 2019). Natalie (chemistry teacher) provides a concise summary of the opportunities to develop skills which research projects provide her students:

> In terms of skills development, research projects are invaluable, students are finding the skills through experience of research, communication skills, teamwork, planning, leadership…and this increases their confidence and self-belief. (Natalie, chemistry teacher)

Both Jacinta (physics teacher) and Melanie (technician) link the opportunities for students to communicate and share their experiences with increased levels of individual confidence, Jacinta said, 'it has been a real privilege to see some students who are not the most confident, take up opportunities to speak about their research and shine' and Melanie shared, 'students involved in research projects become so confident and eloquent and have so many interesting things to say.' Ellen (general science teacher) described how requiring students to communicate their research was, in her opinion, a 'huge skill':

> An important part of the process if that there is a presentation of work, where they share their findings at the end of the project…thinking through how to present their work in an interesting and informative way to adults and to other students is a huge skill, you can know all sorts of interesting things but if you are not able to communicate them then that is problematic. (Ellen, general science teacher)

Anthony (physics teacher) highlighted the specific opportunity that presenting at conferences provided for students to develop their communication skills and that presenting in this context went beyond the expectations and experiences provided in the school environment:

> The pupils are not that confident about standing up and talking, and developing those skills are an important thing as well but whilst we spend a lot of time being busing doing PowerPoints, they are not often actually that good. Having to present research at conferences, standing beside a poster, and having to prepare talking points, that is a very good way to develop communication skills. (Anthony, physics teacher)

Another feature of communication that teachers highlighted as being valuable were the opportunities for students to share their work with the wider school community, perhaps contributing to workshops with younger students or presenting to parents and other interested adults at a school event. John (physics teacher) described how, for academically able students, sharing research with younger pupils gives them opportunities to develop communication skills that went beyond John's own expectations of their abilities. This has caused John to reassess his preconceived ideas about what they were able to achieve:

> Some boys run science clubs with Year 6 students…and these boys here…they are academic, but they still need personal and emotional development, communication and other skills…and these are boys that you wouldn't expect necessarily to be able to talk to primary school kids, these are boys who are really able academically, but not what you would call natural communicators and yet they continually surprise me when I see them explaining particle physics to eight year olds in such a clear and compelling way, and using analogies like lego, ideas that I haven't had and it has taught me not to jump to conclusions about people and to give people opportunities. (John, physics teacher)

John also recognised the ways in which research projects were an enriching cultural experience for overseas students who were living and studying in the UK for the first time and that working with local schools enhanced their understanding of the UK and provided an opportunity for these students to contribute to the wider community:

> The students are going out into the community and contributing and for many of them, who come from significant privilege it is a very important importunity for them to give something back, and for some of our boarders who come from outside the UK it is a great chance for them to get to know the wider context of the place where they live and study, and get a much more rounded experience. (John, physics teacher)

Dean (chemistry teacher) also highlighted the added confidence and self-efficacy that students developed when sharing their research with younger students and supporting others to develop their own experiences of scientific inquiry:

> The skills that students have learnt through research projects, they can then use those to help younger students, supporting students in local primary schools to do science investigations. That not only helps those younger students become enthused with science, but it gives the older students added confidence, that sense of leadership, that they have been able to do into that primary school and deliver something important and valuable. (Dean, chemistry teacher)

In addition to communication skills developed through presenting and sharing their research, teachers also described how working in the context of open-ended investigative project work enabled students to enhance their ability to work as part of an effective team, with varied levels of experience and ages and with students they might not usually work with:

> They are working as a team, with other students, they develop confidence, learning to speak and work with students they otherwise wouldn't interact with, learning to assimilate and communicate ideas to those who might not be at the same level of understanding, they develop mentoring skills and these are all attributes that can be used in life in general, whether you want to be in science or not. (Sarah, technician)

> The skills that the kids learn through this are huge, they learn about team building, getting over disagreements and working together, they have a chat at the beginning of each session and work out what needs to be done and who is going to do it. (Marion, biology teacher)

Unlike school students, engaging in research is an expectation for most undergraduate degree programmes, and a number of studies have demonstrated the importance of undergraduate research opportunities, specifically in STEM subjects to enhance students' skills and university experiences (Graham, Frederick, Byars-Winston, Hunter, & Handelsman, 2013; Hunter, Laursen, & Seymour, 2007; Linn, Palmer, Baranger, Gerard, & Stone, 2015; Thiry, Weston, Laursen, & Hunter, 2012).

Undergraduate research conferences are increasingly part of a student's experience of university education (Walkington, Hill, & Kneale, 2017), with opportunities to present their research at national, university-wide, faculty and disciplinary conferences in the UK, USA, Australia and parts of mainland Europe. Walkington, Hill, and Kneale (2017, p. 416) found that presenting at conferences developed students' confidence, 'giving additional value over and above the recognised benefits of engaging in research.' A key feature of this experience is that the setting has authenticity as a space for genuine research dissemination and is a professional experience for students (Walkington et al., 2017). For undergraduate students, presenting at a research conference is a 'threshold experience' that develops their capacity for self-authorship. Threshold experiences are events in a learner's development where, through participation in a defined event (e.g. presenting at a conference), they move to a more advanced stage of development. In the case of an undergraduate student, the threshold experience of presenting research at a conference can enable individuals to reposition themselves as a researcher as opposed to a student (Meyer, Land, & Baillie, 2010).

More recently, researchers have considered the experiences of high school students who actively participate in academic conferences, sharing their research findings in authentic settings to audiences of their peers, teachers and university-based scientists and researchers (Rushton et al., 2019). In a study which included 27 high school students from 4 high schools in the UK, Rushton et al. (2019) showed that students who participated in academic conferences were overwhelmingly positive about the experience and recognised that this was a novel opportunity where they could engage professionally with the scientific community in a way that validated their contribution. The students suggested that presenting at conferences enabled them to develop some aspects of their communication skills, for example, speaking to large audience in a formal setting and managing their performance anxiety. Every participant said that given the opportunity to present at a conference again they would do so, and many described a sense of pride and accomplishment in the achievement of presenting to an academic audience, in an authentic venue. A key aspect of the positive experience described by students was their recognition that through the experience of the conference they had communicated their research to an audience of scientists, and through this dissemination to the academic community, their efforts had been recognised and validated. Students highlighted that they needed support and guidance in the pre-conference and post-conference phases of research as well as during the conference itself, so that they were able to gain the most from the experience. These broad findings that research projects in general and academic conferences more specifically provide valuable opportunities for students to develop communication skills are consistent with the perspectives of teachers who contributed to this current study. Beyond the more formal environment of an academic conference and the associated presentation formats of poster session and oral presentation, teachers in this current research also highlighted other aspects of communication skill development, including improved teamwork and initiatives to engage the wider community with student-led research projects. These community initiatives frequently develop both students' presentation, performance and communication skills but also cascade enthusiasm for science and inquiry to audiences, including younger students. This cascade effect also extends the networks of all students involved, enabling them to forge greater connections with those engaged in science. These opportunities to develop wider networks and experiences of science research and careers are now considered.

8.4 Providing Opportunities for Students to Develop Wider Professional Networks, Connections and Experiences of Science Research and Careers

The networks and wider connections that students can forge through research was recognised by teachers and technicians as valuable. Gordon (physics teacher) suggested that it was important for students to 'go out, beyond their own school and learn about the research of others' and that through these students are 'connected to wider research group beyond the work they do in school.' Some teachers explicitly linked the opportunity for students to present at academic conferences as an important part of developing wider professional networks and experiences beyond their school content. For example, Keith (chemistry teacher) said:

> I get such a buzz when I see the students going to conferences and holding their own with academics and students at universities…these kids are getting to visit large cities like Manchester, York, Liverpool and this makes a big impact on them. (Keith, chemistry teacher)

As has been discussed in relation to students' development of communication skills, teachers and students acknowledged that academic conferences provide opportunities for student research to be visible to, and valued by, the wider scientific community and that this recognition was important to students. Barbara (chemistry teacher) reflected that this recognition from the scientific community was important to both herself as a teacher and her students:

> For students to be part of a research community in school and are connected with researchers beyond, and to know that their work is recognised and valued and that it could form part of the foundation of further academic research, well it is thrilling for me at my age, I can't imagine what it must be like to be 16 and 17 and know that! (Barbara, chemistry teacher)

As well as students' gaining recognition for their work, teachers also suggested that these authentic experiences of research projects and conferences provided opportunities for students to understand and contribute to the world of science beyond school curricula science:

> We all want to offer students experiences and information that is beyond the curriculum so that they are equipped to go on and be citizens of the world that can contribute to science, rather than people that have achieved a GCSE in science. (Jonny, chemistry teacher)

Joseph (physics teacher) linked the development of skills and self-efficacy to the opportunities students have to forge networks through research that extend beyond their own school, so that students can share their experiences and findings with others and through that better understand that the contribution that they have made is seen by others as valuable: complicated sentence—not sure whether/how to simplify:

> The confidence that research projects have developed, that has been the thing that has surprised me the most, we have asked the students when they have gone to university, they have told us that the skills they have gained from working on research projects have stood them in good stead going ahead. (Joseph, physics teacher)

This idea of forming professional networks increasing confidence and self-efficacy connects back to Keith's (chemistry teacher) observation that students visiting large cities and sharing their research with academics 'makes a big impact on them'. Keith also suggests that the confidence students gain through research is something that moves with them into other educational spaces, 'I honestly believe that the students who do these projects feel better in other lessons and that the confidence they get is transferable.' Sarah (technician) also emphasised the importance of research opportunities for developing student's understanding of science beyond school and especially the careers available in the field of STEM:

> I do the work because I love it and I see the difference that it can make to students' lives and careers, I can see that students who have had these experiences have developed careers in STEM, perhaps when they have not been seen as the most able in the classroom, but in project work they do really well and develop skills that help them at university and beyond. (Sarah, technician)

Teachers and technicians observed that both the professional networks and connections that students develop through research projects, as well as the skills related to practical science and communication, are an important part of developing students' self-efficacy in science. The experience of being part of an authentic science research collaboration provided students with insight into careers in STEM and, in addition to Sarah (technician), teachers including Clare and Barbara (both chemistry teachers) recognised these as vital and often implicit sources of STEM careers education. Barbara (chemistry teacher) said:

> For a university academic to go into some of our schools in the poorer areas of our city, it is really significant because some of those pupils would not otherwise get to meet academics, scientists, to be quite honest, so that gives these pupils something to consider career-wise, what could be on offer. (Barbara, chemistry teacher)

Clare suggests that research projects enabled students to understand and experience science careers for themselves and is therefore more likely to have an impact upon their future choices:

> I'd like to think that the kids subconsciously learn more about a career in science, rather than explicitly, with the research project they are experiencing something that is more like the reality of science and so they are getting a better understanding of what a career in science might be like, so we are not explicitly telling them, they are learning it for themselves and I think that ultimately that is a more powerful way. (Clare, chemistry teacher)

Ellen (general science teacher) described how she hoped that the 'legacy' of running research projects in school was that 'students and teachers are more positive about the value of science and the pathways to science careers for our young people.' As well as this overarching sense that research projects further students' understanding of careers in STEM and support more positive attitudes, teachers and technicians shared how research projects were directly beneficial for students' university applications and future employment:

> Many of our students have used their experience of research projects to support their university applications and are able to talk confidently at interview about their new skills and interests. (Baily, technician)

> Students who are interested in applying to university for medicine and related courses, they say that research projects have really helped them because in their interview they are able to talk about the research they have carried out, beyond their regular lessons and curriculum and universities really like that. (Stephen, psychology teacher)

> Research projects help our students to go on to positive destinations once they leave our high school, whether that be on to college, university or employment, they have so many skills that they can bring from research projects, more than some of their peers. (Anthony, physics teacher)

Teachers and technicians also described how they felt 'better able to give good advice' to students regarding which university course might best suit their skills, experiences and interest and that they had 'a better understanding of what university academics were looking for in applicants' (Bailey, technician). Some teachers highlighted how research projects provided a more equitable opportunity for young people to develop their skills and experiences in STEM, giving more students experience of what studying at university might be like:

> The experiences that you develop in research projects are more equitable than that work experience that is based on who you know, or more importantly, who your parents know and through research projects you get that wider experience of science and education beyond school, your home town and get some experience of university-style learning. (Joan, technician)

As well as providing insight into STEM careers and university courses, some teachers shared the perspective that supporting students to develop new networks and connections could positively challenge their previously held stereotypes about science and scientists and help them to recognise their own capacity as scientists:

> The STEM ambassador we initially worked with was highly motivational, but as an older, white man he perhaps didn't challenge the stereotypical view of who scientists are, so we established a link with a post-graduate, young female archaeologist who came to talk to the students with all her equipment and she explained to the students she was a scientist who solved puzzles about how humans lived in the past, so at the beginning of the project the students had an idea of what an archaeologist looked like and linked them more to history, whereas by the end they saw the science links, and also saw that young women are very much part of science too and I think that will trickle down into their sense that they could be part of science in the future. (Ellen, general science teacher)

In providing opportunities for students to develop wider networks, connections and experiences of science research and careers, research projects could be described as an authentic STEM careers education. Education focused on careers helps young people to better understand the relationship between educational goals and occupational outcomes. STEM university courses are valued for careers such as medicine and engineering (Wang & Degol, 2013), but many students do not know what sort of jobs they could do in STEM beyond these two sub-sets of STEM careers (Bøe & Henriksen, 2013). Career choice is also linked to students' identification of their own ability and confidence levels, as when students are more confident that they can be successful in subjects, including mathematics and science, they are more likely to persist and continue to study STEM subjects (Wigfield & Eccles, 2002). In the context of a career education that has been described as weak (Reiss & Mujtaba, 2017), research has shown that many students make choices about their post-16 education early in their school careers, sometimes as early as 11 years (Tai, Liu, Maltese, & Fan, 2006). Reiss and Mujtaba (2017) highlight the importance of embedding STEM careers education into STEM lessons so that students are aware of the range of career opportunities that are available to them if they continue with STEM post-16 years and the value and transferability of the skills they develop through further study in STEM subjects. Reiss and Mujtaba (2017) also suggest the important role that a parent and/or teacher can have in demonstrating to a young person that STEM subjects are worthwhile, and that the young person is capable of continued success in STEM. Research projects provide the context for young people to understand the range, value and transferability of STEM subjects and careers, and opportunities to genuinely contribute so that they can see that they are already successful in STEM spaces. Beyond the impact that research projects can have on students themselves, teachers described how research projects were a vehicle for students and teachers to make a contribution to wider society through research, and this is the final aspect of contribution considered in this chapter.

8.5 Contributing to Research Efforts that Seek to Address Societal Challenges

Both Peter and Sally (biology teachers) identified that students valued the opportunity to contribute to wider science research, and Peter and Sally attributed this to way that research projects frequently provided students with the opportunity to forge wider connections through the contribution they make. Peter (biology teacher) said: 'they also feel, definitely, that they are contributing to research, there are some of them that regard themselves as researchers having taking part in this project'. Sally agreed, saying: 'the students are part of research community in school, and it was so important to be connected with researchers and make a contribution to that world.' As well as contributing to research, students were also motivated to participate in research projects if they made a connection with the real-life implications of the work and identified with the wider story, Anthony (physics teacher) suggested that, 'students find it easier to engage if they can see the wider relevance of the project.' Peter (biology teacher) expanded upon this idea of 'relevance':

> One of the things that really appealed to the students was that this was applied, it actually has a narrative behind it with real life implications...because it was real they were prepared to sit down and work through the theory and relearn how to apply it. (Peter, biology teacher)

Sarah (technician) also highlighted the value of a project having a strong narrative 'hook' that encouraged students to participate before they understood the intrinsic value of the science that underpinned the research project:

> It is important that projects have relevance for students, whether that is the impact that research makes to society e.g. medical applications, understanding climate change. I often try to find the hook for students, and that is how I can get them enthused and involved at first, through that wider relevance of the project, and the contribution they can make as individuals and then, over time they often develop a love of the science itself. (Sarah, technician)

Keith (chemistry teacher) described how students as a group repeatedly verbally connected their work with making a wider positive contribution:

> The students develop such enthusiasm and confidence, in that they know what they have to do, and they know how to do it and they are able to explain it to people, what they are doing and why they are doing it and they have this little mantra, 'because we are saving the world!' (Keith, chemistry teacher)

Sally (biology teacher) saw research as a successful and positive way of connecting students to problems that impact their lives, now and in the future:

> Through research, students have become enthusiastic about biodiversity and have seen the real benefits that biodiversity brings them…it is that personal link, that tangible link that people have with biodiversity and research has created that link for the students and I think that is not just relevant to our project, a tangible, personal link will get more students involved in science in general because they have had that connection with real research…research can be a mechanism to develop a personal link with a subject for a student. (Sally, biology teacher)

In their previous study involving 17 research-active teachers, Rushton and Reiss (2019) suggested that only a minority of teachers highlighted that research projects provided students with an opportunity to contribute to wider research efforts that contribute to societal development. They contended that this might be an aspect of the experience of being research active that develops over time, or when a teacher is able to spend sustained time over an academic year working 'intensively' on research projects with students. Rushton and Reiss (2019) also contended that it may take time for students to recognise the contributions they are making to research and the wider world, but that this capacity to identify their contribution is possible and extremely valuable in developing confidence, self-belief and in growing and sustaining intrinsic motivation. In this larger study of 53 research-active teachers and technicians, the notion of research projects providing a vehicle for students to make a greater contribution to research community and wider society became more apparent, particularly when it is possible to link research projects to wider environmental concerns, for example, biodiversity loss, global warming and sustainability. Mabel (biology teacher) described how she explicitly framed the research project as an opportunity to contribute to the global challenge of anthropogenic climate change:

> I used experiences from the research project to share in a school assembly about evidence, the importance of making decisions based upon evidence…and then linking that to a bigger problem, like climate change, where students say that they think they can't make a difference, and then I can say to them that they can do something, they can get evidence to provide to scientists so that together you can find solutions to the problem. (Mabel, biology teacher)

As described by Sarah (technician) previously in this chapter, the narrative of sustainability and the environment provided students with the initial hook to participate but also, in the experience of Sophie (biology teacher), gave the students an incentive to persist through periods of challenge:

> I think the sustainability and the importance of the environment and the area that we are working in, that is really important for the students and it is that passion for the environment that really gives them motivation to continue on when it gets difficult or they run into a challenge, they see it as a challenge to overcome rather than a problem that they can't solve or a failure. (Sophie, biology teacher)

As well as research projects providing a pathway for students to develop and harness their passion for the environment, it also enabled teachers to share their own interests and enthusiasm for environment and sustainability issues:

> The climate change narrative and hook does motivate me, originally the group was about getting students more involved in science but now climate change is more and more visible as an issue and some of the students saw

> the research project as a way of making their own contribution…striking was not really an option as to join the climate marches they would have had to travel there by car which kind of defeats the point, and they have written to their MP but they wanted a more consistent and immediate way to be involved. (Mabel, biology teacher)

Teachers and technicians reported that participation in research projects developed some students' sense of agency in the context of global challenges and that this increase in agency was found in school contexts which had previously provided school clubs and projects with an environmental focus:

> Over the years we have done lots of different clubs and projects that have tried to help students learn about their environment and to be responsible citizens, and these have been good for the students but when we got them involved in research we really saw them take the lead with the project and I think that has translated into them having a sense of ownership of the problems their generation faces. (Joshua, biology teacher)

There was also recognition from teachers that environmental challenges, including biodiversity loss and climate change, were often communicated in a negative way to students, and that students discussed with them that being part of research projects gave them a way to contribute in a constructive way:

> Getting young people to think positively about the environment can be a challenge as in their lessons they are getting the accurate message that the vast majority of scientists view climate change as a global challenge, but this can be reported in a very doom-laden way, with lots of shouty headlines. When students were involved in the *MELT* project, they were part of the research team, part of the solution, they had a sense there was a positive way forward and they were part of it. (Hasan, physics teacher)

Teachers and technicians suggested that their students' environmental agency was developed through research projects for some, but not all. Melanie (technician) described how different students responded to the *MELT* research project:

> Some of the students who worked on *MELT* had no interest in the environmental aspect of the project when they started, for them it was an interesting project, a puzzle to get their teeth into and try and solve, but for some of these students, during the course of the project they started to talk about the environmental aspect to me and their peers. I think the consequences and the wider context of what they were studying in an academic way in the research project started to creep into their wider thinking. (Melanie, technician)

Teachers also explicitly described how students who had participated in research projects with an environmental focus were motivated to consider their own behaviour in the context of global warming and biodiversity loss and also identified ways in which these personal changes could be shared with their wider school community. Ellen (general science teacher) described this change:

> The project had only been running in my school for a few months and some of the students who were part of it were in my form and they asked if they could lead an assembly to talk about the project to their year group. When they presented, I was surprised that they included lots of information about biodiversity in their local area, research that they had done independently and they included a section at the end of the assembly that gave the audience some actions they could take to look after their environment e.g. not mowing all of the lawn, creating habitats for bees, checking for hedgehogs before lighting a bonfire but also, reducing energy use. The project really inspired them to think about their actions and then they wanted to share that with their peers. (Ellen, general science teacher)

Marina (general science teacher) also described how research projects supported students to reflect upon their environment-related behaviour:

> The students are able to think about environmental themes that are linked to their project, so bees declining and the importance of pollinators and the wider climate change crisis, so they are able through the project to connect with overarching ideas and that is helping them to make more informed choices about what they do related to the environment. (Marina, general science teacher)

These are clear examples of how, from the perspective of a teacher, students have identified the actions they and their peers can take to change the impact they have on their local and global environments, and how students demonstrate increased agency in communicating these issues with their peers.

As has been outlined in Chap. 3, interviews with teachers and technicians took place during the period December 2017–January 2019. Although the climate strikes were referenced by Mabel (biology teacher) this was the only example of linkage made between student participation in research as a form of environmental activism. However, it should be noted that the interviews with teachers and technicians predominantly took place before the recent upsurge in global action regarding the environment, focused on 'school climate strikes' which were held in March, May and September 2019. This wave of environmental action was inspired by the protest of Swedish teenager Greta Thunberg which began in August 2018. Relatedly, wider research focused on initiatives that involve the public in research, or 'citizen science' has explored the impact of school-based programmes; these are predominately situated in ecological and environmental studies, reflecting the early trend in citizen science programmes as a whole (Bonney et al., 2009). School-based citizen science projects with an environmental and ecological focus described in the research literature have a clear overlap with the research projects teachers and technicians were involved in when interviewed for this current study. Therefore, the broad learning from school-based citizen science literature can provide useful further insights into the impacts on students who participate in ecological and environmental research projects with their teachers and other external partners. For example, ecological and environmental studies have considerations of place and agency at the centre of their research, which means they provide useful settings in which to consider student agency (Ballard, Dixon, & Harris, 2017). This is important when considering, for example, whether school-based citizen science programmes provide young people with an opportunity to make a genuine contribution to scientific knowledge (Ottinger, 2010) or to lead a research project (Calabrese Barton, 2012). Ballard et al. (2017) and Harris and Ballard (2018) found that when students are given ownership of data collection and analysis, disseminating and communicating research findings, and providing guidance to their local community, young people are more likely to develop agency as they are positioned as experts. Ballard et al. (2017) also recognised that students develop agency when they perceive that the research project they are contributing to is authentic, particularly when their work is associated with multiple aspects of the project work and is not solely linked to data collection that is then used by professional scientists. School-based citizen science projects that develop student agency include repeated experiences for students that generate or develop connections to a place and provide students with explicit ways to contribute to authentic research science (Ballard et al., 2017). Sharing their findings with outside audiences is important for students to develop their identity as researchers, providing their expertise to a range of stakeholders who recognise the value of their work (Ballard et al., 2017; Harris & Ballard, 2018). As has been shown in this chapter, teachers and technicians have highlighted how research projects provided opportunities for students to develop their agency in the ways they behave and respond to issues focused on the environment and sustainability. Furthermore, research projects provided opportunities for students to gain practical and communication skills, build connections and experiences of STEM courses and careers that enable enhanced self-efficacy. Teachers and technicians increasingly acknowledged that students valued the opportunity to better understand global challenges, and being part of research that seeks to provide solutions is a key source of motivation to initially participate and to continue through periods of challenge and difficulty.

Over the course of this, and the previous four chapters, I have considered in detail the experiences of science teachers and technicians who are research active with their school students, exploring themes of freedom to teach, (re)connection with science/research, collaboration, professional development and student/societal development through research. In the following chapter, I synthesise these themes to consider the ways in which, through the experience of being research active, teachers and technicians develop a new model of professional identity, namely as 'teacher scientists'.

References

Ballard, H. L., Dixon, C. G., & Harris, E. M. (2017). Youth-focused citizen science: Examining the role of environmental science learning and agency for conservation. *Biological Conservation, 208*, 65–75.

Bennett, J., Dunlop, L., Knox, K. J., Reiss, M. J., & Torrance Jenkins, R. (2018). Practical Independent Research Projects in science: A synthesis and evaluation of the evidence of impact on high school students. *International Journal of Science Education, 40*(14), 1755–1773.

Bøe, M. V., & Henriksen, E. K. (2013). Love it or leave it: Norwegian students' motivations and expectations for post-compulsory physics. *Science Education, 97*(4), 550–573.

Bonney, R., Cooper, C. B., Dickinson, J., Kelling, S., Phillips, T., Rosenberg, K. V., et al. (2009). Citizen science: A developing tool for expanding science knowledge and scientific literacy. *BioScience, 59*(11), 977–984.

Calabrese Barton, A. M. (2012). Citizen (s') science. A response to "The Future of Citizen Science". *Democracy & Education, 20*(2), 1–4. Retrieved from https://democracyeducationjournal.org/cgi/viewcontent.cgi?referer=https://scholar.google.co.uk/&httpsredir=1&article=1044&context=home

Dunlop, L., Knox, K. J., Bennett, J. M., Reiss, M., & Torrance Jenkins, R. (2019). Students becoming researchers. *School Science Review, 100*, 85–91.

Graham, M. J., Frederick, J., Byars-Winston, A., Hunter, A.-B., & Handelsman, J. (2013). Increasing persistence of college students in STEM. *Science, 341*(6153), 1455–1456.

Harris, E., & Ballard, H. (2018). Real science in the palm of your hand. *Science and Children, 55*(8), 31–37.

Hatfield, P. (2010). Using line intensity ratios to determine the geometry of plasma in stars via their apparent areas. *High Energy Density Physics, 6*(3), 301–304.

Hunter, A., Laursen, S. L., & Seymour, E. (2007). Becoming a scientist: The role of undergraduate research in students' cognitive, personal, and professional development. *Science Education, 91*(1), 36–74.

Linn, M. C., Palmer, E., Baranger, A., Gerard, E., & Stone, E. (2015). Undergraduate research experiences: Impacts and opportunities. *Science, 347*(6222), 1261757.

Meyer, J. H. F., Land, R., & Baillie, C. (Eds.). (2010). *Threshold concepts and transformational learning*. Sense.

Ottinger, G. (2010). Buckets of resistance: Standards and the effectiveness of citizen science. *Science, Technology, & Human Values, 35*(2), 244–270.

Parker, B., Fox, E., & Rushton, E. A. C. (2018). IRIS – Promoting young peoples' participation and attainment in STEM and reigniting teachers' passion for science education. *Impact 2*. Retrieved from https://impact.chartered.college/article/parker-iris-stem-students-teachersparticipation-research/

Reiss, M. J., & Mujtaba, T. (2017). Should we embed careers education in STEM lessons? *The Curriculum Journal, 28*(1), 137–150.

Rushton, E. A. C., Charters, L., & Reiss, M. J. (2019). The experiences of active participation in academic conferences for high school science students. *Research in Science and Technological Education*. https://doi.org/10.1080/02635143.2019.1657395

Rushton, E. A. C., & Reiss, M. J. (2019). From science teacher to 'teacher scientist': Exploring the experiences of research-active science teachers in the UK. *International Journal of Science Education, 41*(11), 1541–1561.

Tai, R. H., Liu, A. C. Q., Maltese, D. V., & Fan, X. (2006). Planning early for careers in science. *Science, 312*(5777), 1143–1144.

Thiry, H., Weston, T. J., Laursen, S. L., & Hunter, A. B. (2012). The benefits of multi-year research experiences: Differences in novice and experienced students' reported gains from undergraduate research. *CBE—Life Sciences Education, 11*(3), 260–272.

Walkington, H., Hill, J., & Kneale, P. E. (2017). Reciprocal elucidation: A student-led pedagogy in multidisciplinary undergraduate research conferences. *Higher Education Research & Development, 36*(2), 416–429.

Wang, M. T., & Degol, J. (2013). Motivational pathways to STEM career choices: Using expectancy–value perspective to understand individual and gender differences in STEM fields. *Developmental Review, 33*(4), 304–340.

Whyntie, T., & Harrison, M. A. (2014). Simulation and analysis of the LUCID experiment in the Low Earth Orbit radiation environment. *Journal of Physics: Conference Series, 513*(2), p. 022038. Bristol, UK: IOP Publishing.

Whyntie, T., & Harrison, M. A. (2015). Full simulation of the LUCID experiment in the Low Earth Orbit radiation environment. *Journal of Instrumentation, 10*(03), C03043.

Wigfield, A., & Eccles, J. S. (2002). The development of competence beliefs, expectancies for success, and achievement values from childhood through adolescence. In A. Wigfield & J. S. Eccles (Eds.), *Development of achievement motivation* (pp. 91–120). Academic Press.

9

A Model of the Teacher Scientist Identity

9.1 Introduction

This chapter brings the wider science education literature discussed in Chaps. 1 and 2, together with empirical findings from this research shared in Chaps. 4 through to 8 inclusive, and proposes a model of science teacher identity that teachers develop in the context of engagement with and participation in authentic research projects. This model of teacher scientist identity has three distinct but related facets: (1) inquiry identity, (2) subject identity and (3) social justice identity. At the outset, each of the three facets is briefly considered in the context of the relevant literature to better understand what these might mean in the context of a teacher's identity. This discussion begins with exploring the facet of 'inquiry identity'.

E. A. C. Rushton, *Science Education and Teacher Professional Development*, Palgrave Studies in Alternative Education, https://doi.org/10.1007/978-3-030-64107-8_9

9.2 Inquiry Identity

As discussed in Chap. 2, research that explores science teacher identity in the context of inquiry has predominantly considered the experiences of preservice teachers and has focused on their teaching practices, as opposed to their identity development. A fuller understanding what an inquiry identity might look like in teachers who are experienced in inquiry is a key aim of this research and this is discussed later in this chapter. Thus far, the research suggests that an inquiry identity is broadly focused around two areas which Varelas, House, and Wenzel (2005) describe as (1) *science as practice* and (2) *a community of practice.*

Science as practice is centred on experiences where teachers can contribute to knowledge and further understanding through engaging with the complexity of theory and data in an authentic way (Rushton & Reiss, 2019; Varelas et al., 2005). The aspect of communities of practice highlights the collaborative nature of science (Bryce, Wilmes, & Bellino, 2016), where individuals have freedom and autonomy whilst also mentoring and supporting others (Varelas et al., 2005; Ufnar & Shepherd, 2019) in online and offline contexts (Bang & Luft, 2016). Drawing on the research of Eick and Reed (2002), it is important to note that preservice teachers who identified themselves as 'constructors of knowledge' as opposed to 'tradition learners' were more able to implement structured inquiry as part of their school placement. This identity as a 'constructor of knowledge' is perhaps more aligned with science as practice, giving an individual agency to seek to further understanding rather than be in receipt of knowledge.

Relatedly, Bang and Luft (2016) highlight how teachers whose identity is positioned as a 'learner-of-being-a-teacher' in the context of inquiry-based teaching are able to explore and develop multifaceted identities that draw on a range of sources, contexts and experiences rather than that rooted in a single ideology, approach or curriculum. Rushton and Reiss (2019) also emphasise the value of opportunities for teachers to develop multifaceted professional identities that provide rich experiences and sustain through periods of challenge. The science as practice aspect of an inquiry identity could shape this facet of a science teachers' identity as both 'a-learner-of-science' and 'a-teacher-of-science', enabling teachers to construct multiple professional identities. This is consistent with Melville, Bartley, and Fazio (2013) who suggest that preservice teachers should develop dual identities as students of science and preservice teachers of science. However, I argue that this approach to professional identity development should continue throughout a teachers' career and that engaging with both science as practice and the community of practice aspects of inquiry identity, as a teacher scientist, makes this possible throughout a career. As part of research with high school teachers (based in England) who implemented open-ended investigations with students during an A-level course (post-16 years), Dunlop, Turkenburg-van Diepen, Knox, and Bennett (2020) identified six different ways that teachers perceived open-ended investigation which corresponded to the different emphases that teachers had for student learning, including the 'teacher-inquirer'. The 'teacher-inquirer' is a teacher who is focused on 'real science' and curiosity about the world, and the direction of the research topic and the choice to include external partners is motivated by the interests and ideas of their students. Dunlop et al. (2020) suggest that the 'teacher-inquirer' has a role focused on facilitating student involvement rather than directly participating in the research project. In their research, Dunlop et al. (2020) were focused on teachers' perceptions of learning outcomes rather than the ways that facilitating open-ended investigations shaped teachers' identities; however, both the concept of 'teacher-inquirer' and the fact of inquiry identity highly value a space where teachers and students are curious and seek to make sense of the world by asking questions and iteratively exploring answers.

Based upon research that considers inquiry identity, fostering such an identity depends on individuals enacting science as practice, where they are able to contribute to knowledge and further understanding in the context of a supportive and collaborative community of practice. As is also noted in both the social justice and subject facets of identity, developing and maintaining an inquiry identity can be challenging. Tensions can occur when identities that are rooted in science inquiry are brought into different, if related, contexts (e.g. classroom). Varelas et al. (2005) and Dreon (2008) also recognised the anxiety that preservice teachers frequently experience when implementing inquiry-based pedagogies in the classroom. Sustaining an inquiry identity is made possible when individuals are part of a supportive community of practice (Bang & Luft, 2016; Rushton & Reiss, 2019, 2020; Ufnar & Shepherd, 2019). Closely related to the inquiry facet of a teacher, scientist identity (in particular the 'science as practice' element) is subject identity, and this is explored in the following section.

9.3 Subject Identity

Both qualified and preservice teacher identities have long been recognised (but not universally e.g. Nieswandt, Barrett, & McEneaney, 2013) as being most closely aligned with their subject identity, as opposed to an identity of pedagogical experts (Beijaard, Verloop, & Vermunt, 2000). Beijaard et al. (2000) suggest that this is especially true for early career teachers, science and mathematics teachers and male teachers. However, what comprises a subject identity? What are the norms associated with a subject identity as opposed to a teaching identity?

Research suggests that subject identity is rooted in an individual's self-efficacy in relation to their subject knowledge (Beijaard et al., 2000). An individual's capacity as a 'subject knowledge expert', who has a deep and extensive understanding of the subject area, is a crucial aspect of science subject identity, and subject knowledge has been found to be a core part of establishing positive professional identities (Beijaard et al., 2000; Irving-Bell, 2018). Furthermore, research focused on preservice teachers has found that subject identity is an aspect of professional identity that is well developed at the outset of teachers' careers (Manning, 2017; Woolhouse & Cochrane, 2015), and Rushton and Reiss (2019) argue that early career teachers have not yet had time to develop identities that are significantly grounded in aspects beyond their subject. In research that considers teachers' perceptions and emphases for student learning through open-ended investigations, Dunlop et al. (2020) describe one of six perceptions as, 'the teacher-scientist' where the learning focus is on, 'the state of the field' within science, which is frequently chosen and defined by the teacher and involves collaboration with external partners over several years and leads to outputs including presentations and journal articles. Although identity is not the lens through which Dunlop et al. (2020) explore the ways teachers perceive, interpret and teach open-ended investigations, the perception of the 'teacher scientist', with a focus on the 'state of the field', is arguably aligned to the facet of subject identity which is grounded in teachers' belief and perceptions of the value of continually developing as subject knowledge experts.

As well as subject knowledge, El Nagdi, Leammukda, and Roehrig (2018) suggest that a subject or science identity rests upon an individual's ability to apply specific knowledge (e.g. engineering) to solve real-world problems for the benefit of society, and that teaching approaches that are aligned with a science or subject identity include inquiry-based, problem-based or project-based learning (El Nagdi et al., 2018). Research that considers the experiences of science teachers who are career changers, that is, those who move into teaching from careers including scientific research, engineering and industry, also highlights the importance of professional identities that are rooted in science and subject identities drawn from individual's prior experiences (Antink-Meyer & Brown, 2017; Molander & Hamza, 2018; Snyder, Oliveira, & Paska, 2013). The facets of subject identity and inquiry identity are clearly linked: inquiry, experimentation and discovery are the ways in which new science subject knowledge is created and understanding furthered. The facet of social justice is rooted in the type of science teacher an individual is or seeks to become, the way they understand the purpose and value of science education and the contribution they wish to make to society. This, the third facet of a teacher scientist identity is now considered.

9.4 Social Justice Identity

When developing a social justice identity as an educator, and specifically a science educator, emphasis is placed upon actively developing and promoting caring and equitable approaches to learning, for all (Rivera Maulucci, 2013). Social justice identities are rooted in providing greater access and agency to those who would normally be marginalised, so that (science) education becomes more pluralistic and representative of the society it seeks to serve (Luehmann, 2016; Rivera Maulucci, 2013).

Socially just science education acknowledges the 'struggle' and power imbalance that is present and recognises the role of teachers and informal educators in challenging and reforming power relations and ways of knowing so that science fully represents and benefits from the knowledge, cultures and perspectives of society as a whole. Both formal and informal spaces of learning are recognised as important spaces for both developing and supporting an educator's social justice identity and in helping to promote equity across different levels in society, for example in the home, classroom, school and community (Luehmann, 2016; Rivera Maulucci, 2013).

Dawson (2017) highlights the importance of understanding social justice as both *redistributive*—how resources can be distributed most equally between social groups (Rawls, 1971)—and *relational*—how resources can be distributed most equitably through valuing individual difference (Young, 1990). Dawson (2017) combines both redistributive and relational social justice as a basis for framing equity and inclusion in out-of-school science learning, using the three concepts: *infrastructure access*, *literacies* and *community acceptance*. Firstly, *infrastructure access* or the extent to which people can access and shape spaces and resources. Secondly, *literacies* or whether languages and ways of knowing are pluralistic and determined by participants as well as providers. Lastly, *community acceptance*—the ways new participants are welcomed into spaces of learning and have their experiences recognised and valued. These three concepts are also levels of analysis that are present to a greater or lesser extent depending on how socially just the context or approach is (Dawson, 2014; Grabill, 1998; Porter, 1998).

In addition to the teacher perspectives of open-ended investigations described by Dunlop et al. (2020) as 'the teacher-scientist' and 'the teacher-inquirer', a further four perspectives were identified through this research including 'the instrumentalist', 'the scaffolder', 'the independence-builder' and 'the personal developer'. Taking an 'instrumentalist' perspective' positions open-ended investigations as being a pathway or tool to achieve a tangible outcome (e.g. award) and/or recognition for the student. In contrast, the focus of a 'scaffolder' is on the interaction between the student and teacher which is motivated to bring about students mastery of scientific techniques (e.g. laboratory methods or software applications) so that the student becomes increasing independent over time. Relatedly, 'the independence-builder' focuses on developing students' independence, providing space and time for students to make mistakes and learn from them. The key difference with 'the scaffolder' is that science is not the focus for the 'independence-builder', whereas the perspective of the 'scaffolder' is grounded in 'real science'. Lastly, 'the personal developer' perspective is grounded in the students' holistic development where, through participation in open-ended scientific investigation where they encounter and overcome difficulty, students develop resilience and confidence. To be clear, Dunlop et al. (2020) described each of these four perspectives through a focus on learning, rather than notions of student or teacher identity. However, the focus on student outcomes that include recognition and affirmation of students, increasing independence and personal growth are perceptions of open-ended investigations which are akin to ideas of relational social justice. For example, through perceptions of 'the scaffolder', teachers enable students to gain infrastructure access to spaces of science, and through perceptions of 'the instrumentalist', teachers enable students to develop science literacies; finally, students are supported through perceptions of 'the independence-builder' and 'the personal developer' to have the confidence and resilience necessary for community acceptance.

Drawing on these social justice literatures, we can understand that developing a social justice identity as an educator means recognising and proactively challenging inequity; it requires care and empathy and a desire to develop agency in others; it is rooted in the perspective that people's differences are valuable and their individual needs matter. As researchers have previously highlighted, developing a social justice identity is a demanding professional endeavour (Luehmann & Markowitz, 2007; Richmond, 2016), and educators need access to non-judgemental, supportive spaces where they can freely discuss, experiment and explore issues of power, equity and social justice (Adams & Gupta, 2017; Luehmann, 2016; McIntyre & Hobson, 2016).

Having outlined how the three facets of inquiry, subject and social justice forms part of a teacher scientist identity based upon the wider science education literature, these facets are considered further in the context of the empirical data discussed in Chaps. 4, 5, 6, 7 and 8.

9.5 Further Understanding a Model of Teacher Scientist Identity

Each of Chaps. 4–8 presents detailed analysis of one of five themes that capture the experiences of science teachers who have participated in sustained, authentic research with their students. I argue that each of these themes coalesce around the three facets of a teacher scientist identity. Table 9.1 sets out the way the themes from this empirical research can be situated within the three facets of identity.

Figure 9.1 suggests how these three facets come together to form a model of teacher scientist identity. A Venn diagram has been used to present this model to underline the fluid nature of the facets and the themes; there will always be some overlap between them, and different facets will feature to a greater or lesser extent depending on the individual concerned. This model of teacher scientist identity draws on the *Social Identity Approach* (SIA) (discussed in Chap. 2) as I argue that the five Is of *identification, ideation, interaction, influence* and *ideology* (Haslam, 2017) provide the mechanisms through which these three facets are developed (Fig. 9.1). In the following discussion, the facets of the teacher scientist identity and the ways in which they develop and interact are explored drawing on the empirical findings presented in Chaps. 4–8.

9.5.1 Inquiry Identity and the Teacher Scientist Model

As has been discussed above, the inquiry facet of teacher scientist identity is focused on science as practice, where teachers can make a genuine

Table 9.1 The facets of a teacher scientist identity drawn from the superordinate themes

Facet of identity	Themes from Chaps. 4–8
Inquiry	(B) (Re)connection with science/research (C) Collaboration
Subject	(A) Freedom to teach (D) Professional development
Social justice	(E) Societal impact/student impact

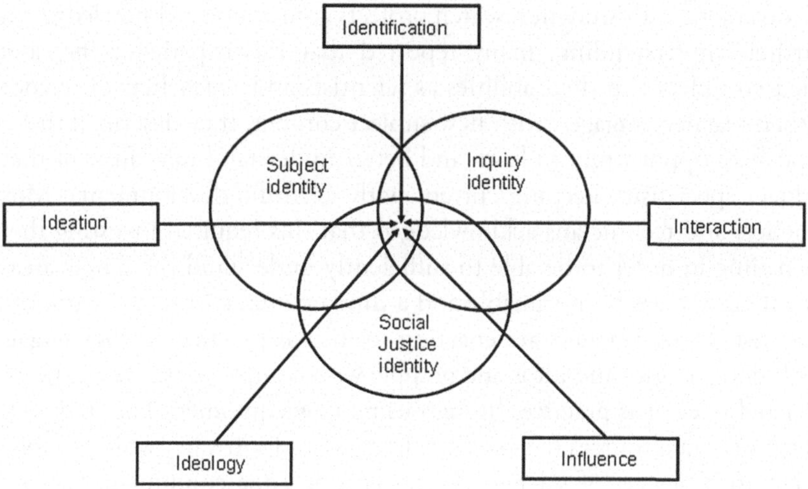

Fig. 9.1 The model of the teacher scientist identity

contribution to knowledge, and involves teachers in a wider community of practice. As presented in Table 3.1, the two themes from this empirical study that most relate to inquiry identity are *(re)connection with science/research* (which I argue documents teachers' engagement with science as practice) and *collaboration*, or teachers' participation in communities of practice, and each of these aspects are now discussed in turn.

As has been explored in Chap. 5, teachers who participated in authentic research projects with school students shared that this experience had supported their fundamental connection to scientific inquiry and for some, had enhanced their sense of being a scientist as opposed to a teacher of science. This sense of (re)connection was apparent for those teachers with and without a PhD in a STEM field and was developed in three ways: (1) being part of discovery, (2) engaging with new subject content and (3) using new equipment and incorporating more practical work. The experience of 'being part of discovery' was something that enthused and engaged teachers with science. Teachers reported that they enjoyed the experimental and cutting-edge nature of science research projects and felt that this work was both novel and important. In being part of discovery, teachers are engaging with science as practice, where their involvement with student research projects contributes to knowledge and furthers understanding; many reported that this experience increased their confidence in their abilities as scientists and as teachers of science. When teachers engaged with new subject content, they described this as a positive opportunity to learn and better understand new areas of their subject specialism, keeping current with scientific developments. Most teachers and technicians acknowledged that this required investing their own time in order to be able to sufficiently understand these new areas, but that this was both valuable and a fundamental part of the work of a scientist. Here, teachers are enacting science as practice, as they engage with new theories and ideas and grapple with complex data. Teachers also enacted science as practice through using new equipment (e.g. radiation detector), data sources (e.g. satellite images) and software, which enabled them to 'feel part of science'. Using cutting-edge equipment that was authentic, in that it was also used by laboratory-based research scientists, further supported teachers' identification with science as practice, prompting them to promote inquiry and experimentation with open research questions as part of authentic, 'real world' investigations. In this sense of being part of discovery, engaging with new subject content and use of new equipment, teachers and technicians are (re)connecting with science as practice. Drawing on the work of Haslam (2017), I argue that teachers enact science as practice through their participation in research projects broadly through the mechanism of *identification* (Fig. 9.1), that is, teacher involvement in science research projects enables them to identify with science as practice in a way that supports their professional identity. However, the process of Identification, as with all aspects of the Social Identity Approach, is a corporate rather than an individual endeavour. In addition to science as practice, collaboration or communities of practice is an aspect of the inquiry facet of a teacher scientist identity and this is now explored.

The focus of Chap. 6 is collaboration, and the discussion of teacher experiences identified three distinct ways of collaboration including (1) collaboration as a new and different way of working with students, (2) working with external partners including scientists, university staff and teachers and students from others schools and (3) establishing and developing collaborative networks. I suggest that this experience of collaboration is grounded in being part of a community of practice, where people have freedom and autonomy within a supportive community that develops networks in a range of contexts to enable sharing of ideas and mutual encouragement. Teachers and technicians who shared their experiences as part of this research highlighted that research projects encouraged student autonomy and independence and that teachers and technicians had different roles from those that they usually enacted with students. These new and different roles were described using a variety of terms including advisor, coach, facilitator and mentor, and some teachers suggested that their role developed over time from one that was more didactic to that which was more subsidiary or peripheral, as students took on greater ownership of the research. Students also developed new roles and ways of collaborating, working in peer and near-peer mentoring roles as part of a vertical research group that drew students from across multiple year groups and continued over successive years. Both staff and students valued opportunities to collaborate with external partners, whether these were based in research institutions or other schools and community groups. Teachers and technicians described how these external collaborations provided practical and intellectual support as well as giving the projects additional authenticity and validity with students. These shared experiences also highlighted how teachers and technicians were able to build and extend their networks, or communities of practice, incorporating opportunities to widen their pool of collaborators through in-person and online events. In these ways, research projects provide teachers and technicians with the context and tools to build communities of practice that engender and enhance their identification with science and practice. Through collaboration, teachers are drawing on the mechanism of *interaction* (Fig. 9.1) which Haslam (2017) argues is what develops and galvanises social identities, shapes the extent to which individuals feel part of a group (or community of practice) and is the central part of why social identities have a positive impact on professional identities.

9.6 Subject Identity and the Teacher Scientist Model

An individual's confidence in the strength and depth of their subject knowledge has been shown to be a critical part of a positive subject identity for (science) teachers. In addition, subject identity is also built upon the ability of an individual to apply their subject knowledge to authentic, real-world problems and provide solutions. As has been discussed as part of the facet of inquiry identity, research provides teachers with the chance to (re)connect with science and research in a way that further develops their subject knowledge. Furthermore, this (re)connection to science and research is an important contributor to strengthening a positive subject identity. As part of this current study, the two empirical themes of 'freedom to teach' and professional development are explored in Chaps. 4 and 7, respectively, and provide further insight as to the facet of subject identity.

When discussing 'freedom', teachers and technicians described three aspects: (1) the flexibility of research projects as an approach to science education, (2) freedom from the constraints of external examinations and curricula and (3) the variety of teaching methods and approaches that were available when supporting school-based science research projects. Research enabled teachers and technicians the freedom to develop their subject knowledge and that of their students in a flexible way that is rooted in experimental work, with the opportunity to work in a way where trial and error and 'mistakes' are seen as a positive part of innovation and discovery. By participating in research projects that were not linked to external exams or curricula, staff and students had the freedom to work beyond the necessarily prescribed content, disciplinary boundaries and methods of assessment, particularly those linked to practical work and real-world inquiry. Research projects enabled teachers and technicians to teach and develop subject knowledge in a variety of ways that were frequently student-led and were situated in real-world contexts, beyond the classroom. In this way, research projects provide an opportunity not only to expand subject knowledge but also to do so in a way that is authentic and occurs collaboratively, in different phases, and over an extended period.

Chapter 7 shares the ways in which teachers and technicians experienced and understood research projects as a form of professional development and identified three distinct aspects: (1) the development of new or enhanced skills, including practical and interpersonal, and subject knowledge, (2) increased recognition of their role and their subject including the development of alternative pathways to professional development beyond traditional management routes, and (3) research projects as a pedagogical approach to science education. When making and creating the facet of subject identity, this first aspect from the theme of professional development is perhaps the most relevant. Teachers and technicians shared how research projects enabled them to enhance subject knowledge alongside the development and/or enhancement of skills. Practical skills included using high-quality microscopes, novel laboratory techniques and computer-based skills, for example, using new software and programming. These skills were developed as part of engaging with and supporting students to access areas of a discipline that were either new to the student and teacher and/or were at the cutting-edge of current understanding about an aspect of science. Teachers and technicians described how research projects enabled them to refresh their subject knowledge by engaging with new theories and debates, and that this approach to maintain a rich subject knowledge was undertaken as an individual teacher or across an entire school science department. As well as generating further practical skills and subject knowledge, teachers and technicians shared how research projects led to their own personal development, increasing their confidence, self-belief and their ability to be reflective practitioners. I argue that this perceived increase in the self-efficacy of teachers and technicians is rooted in the strengthening of their subject identity by the generation and application of greater subject-specific expertise. The recognition and validation that teachers and technicians experience from their colleagues and senior leaders further enhance an individual's sense of being a 'subject knowledge expert'. In developing a subject identity as part of the teacher scientist model (Fig. 9.1), teachers and technicians draw on both the concepts of *identification* and *ideation* (Haslam, 2017). As has been noted when discussing the facet of inquiry identity, identification is a mechanism that enables teachers and technicians to enact science as practice. In doing so, I suggest that this maintains and extends robust and healthy subject identities through encountering new knowledge in the context of opportunities to actively engage with, and shape new understandings and insights in the field of science. Ideation refers to *what* people identify with, and Haslam (2017) suggests that what people identify with is as important as mutual identification. Teachers and technicians describe how they identify with the practice of science, which is part of an inquiry identity, but it is also part of a subject identity as the practice of science underpins and generates scientific theories and understanding. Through ideation, teachers and technicians enfold themselves in their subject identity which has a set of group norms that both describe what people in this subject group do and what they should do. This is seen in the experiences shared by some teachers and technicians when they describe furthering subject knowledge, whether through research projects or not, as something that is to be expected of science teachers.

9.7 Social Justice Identity and the Teacher Scientist Model

When thinking about social justice in the context of the role of an educator, the literature highlights the importance of educators being able to recognise and then to challenge inequity in their context. Socially just educators are found to have empathy and seek to develop agency in the children and young people they work with. Moving beyond redistributive social justice, educators both recognise and value the differences and distinct needs of each individual. As part of a reflective exercise focused on their professional development, one teacher explicitly discussed social justice and equity in the context of her participation in research projects (Chap. 7). Jacey (psychology teacher) highlighted the way research projects enabled her to provide opportunities in science for a more diverse group of students who might not otherwise have such experiences. Jacey was the only teacher who explicitly discussed research projects as a way to challenge inequity; however, as explored in Chap. 8, teachers and technicians frequently described how research projects provided ways to develop both the inquiry and communication skills of their students and gave students opportunities to make wider connections, establish networks and gain experience of science research and careers, and these are now explored through the lens of social justice.

Teachers and technicians demonstrated empathy towards their students in their implementation of research projects; indeed, many were initially motivated to voluntarily incorporate research projects as part of their extra-curricular offer because they recognised and responded to the requests of their students. Teachers and technicians sought to provide spaces for research projects to reach as many students as possible, for example holding sessions during lunch breaks and at the end of the school day, supporting a cross-curricular approach to research to include students that they might otherwise not teach, as well as personally inviting and encouraging students who they felt would benefit most and who might not otherwise participate without this prompt. Some recognised that research projects provided enriching cultural experiences for students either through travel to new and culturally different places (e.g. countries) and/or spaces (e.g. research laboratories, online seminars with scientists). A recurring feature in the reflections of teachers and technicians was the emotional support that students required to negotiate both the challenges of research experience as part of their work based in their own school context, and to participate in opportunities beyond more familiar settings. In providing students with logistical support (e.g. organising transport to and from an event) and emotional support (e.g. coaching a student through presentation preparation), teachers and technicians were empathetic in their approach. Some teachers explicitly shared their own thoughts and fears about different phases of research, providing an environment where discussing uncertainties and anxieties was encouraged and where, through collaborative and empathetic working, solutions could be found together.

Teachers and technicians reflected that research projects were a way of supporting students' skills such as inquiry and communication, and their descriptions often feature examples of increased student agency. In thinking about agency, it is helpful to draw on the concept as understood by

Pantić and Florian (2015) where individuals with agency have intentionality or a sense of *purpose*, the *competency* to achieve such purpose and a degree of *autonomy* to act. For example, teachers and technicians described the sense of pride they felt in observing students who, at the outset of a research project, displayed less self-confidence than their peers, participated in opportunities to communicate their research towards the end of their contribution to the project. In their observations and reflections, teachers shared how research projects gave students:

1. a sense of purpose and a framework,
2. opportunities to develop the necessary competencies, skills and experiences to contribute to research projects and
3. participating as part of a research team, with freedom and autonomy to act.

This autonomy was particularly visible in the way students shared their research findings in settings beyond their own schools, where teachers found that students were able to perform to a standard that went beyond the initial expectations of both the teachers and themselves. Here, communicating the findings of research, for example through conferences and visits to local primary schools and in written articles, provided an agentic vehicle for students practice science competently and with autonomy.

Research projects also helped teachers to develop wider networks and experiences of science careers for their students that promoted student agency. For example, when students participated in research conferences, they were able to build networks with scientists from the position of being a contributing member of the wider science community, thus developing both their competency as scientists and their autonomy in becoming part of a community of practice. Furthermore, many teachers observed that contributing to conferences underlined the wider purpose of participating in research projects for many students, and that following this experience they redoubled their research efforts, aided by an increased sense of self-efficacy. Self-efficacy was observed as being derived from students being able to make a contribution to research that, through participation in conferences (both as an individual and as part of a collaborative effort), was seen and acknowledged by members of the scientific community as valuable. Through research projects, students were able to experience science in a real-world context and gain insight as to the career options within science and the range of pathways to those careers. By providing opportunities for students to understand and experience future employment in science in ways that promote agency, there is the potential that science itself may benefit from contributions drawn from a greater diversity of human capital. Beyond individual student development, the experiences shared by teachers and technicians as documented in Chap. 8, research projects provided a pathway for young people to provide solutions to problems that impact the wider world, including disease, biodiversity loss and climate change. These challenges provided a compelling narrative hook for students, who were motivated to make a difference to wider society through their participation in research projects as part of a wider community of practice.

As with the facets of inquiry and subject identity, I contend that when developing and enacting their social justice identity, teachers and technicians draw on the mechanisms from the five *I*s, in this case *influence* and *ideology* (Fig. 9.1), grounded in the research of Haslam (2017). Influence is the extent to which an individual's attitudes, intentions and behaviour are shaped by their group membership, in this case, how much teachers, technicians and school students socially identify with the wider community of science. When students' experiences of research are positive and their contributions are valued and affirmed by external stakeholders, they are more likely to be influenced and share a social identity as a scientist (Haslam, 2017). In seeking to develop opportunities for their students to develop agency in their contribution to science research, teachers draw on an ideology of education that is rooted in social justice and that this ideology provides the wider context for the mechanisms of identification, ideation and interaction (Haslam, 2017). Haslam, Reicher, and Platow (2011) note that it is important to cultivate a range of identities rooted in a shared ideology, with education communities identifying with both superordinate groups (e.g. community of science practice) and subgroups (e.g. school students, teachers, university-based researchers) to reflect and represent diversity and realising this shared identity through activities, for example, research conferences.

9.8 Conclusion

Over the course of this chapter, I have explored in detail the three facets of a teacher scientist model of identity and highlighted how the mechanisms of *identification*, *ideation*, *interaction*, *influence* and *ideology* enable individuals to develop their identity as a teacher scientist. Through the lens of the SIA outlined in Chap. 2, I have emphasised the crucial role that shared identity and group membership play in the identity development of high school science teachers who are research active. I suggest that this model of identity is also relevant to teachers of other disciplines as the facets of subject and social justice identity can be easily translated across to any high school subjects. So too, I argue is the facet of inquiry identity. For example, a geography teacher might view the facet of inquiry identity from the perspectives of place, space and interdependence whereas an English teacher might draw on creativity and imagination as aspects of an inquiry identity. I argue that inquiry, or the act to discover more information, is a core facet of a model of teacher identity that is based in experiences of authentic research and/or creative practice, irrespective of the subject or discipline. The question remains, how might teachers develop their professional identities in this way? What types of activities can teachers participate in as part of their professional development that is grounded in research? How might these develop their practice and therefore their professional identities as teacher scientists or teacher artists or teacher historians? In the final chapter, I outline ten salient practices to support teacher professional development in this way, building on the research of Walkington and Rushton (2019) and drawing on the experiences and insights shared by the 53 key informants as part of this current research.

References

Adams, J. D., & Gupta, P. (2017). Informal science institutions and learning to teach: An examination of identity, agency, and affordances. *Journal of Research in Science Teaching, 54*(1), 121–138.

Antink-Meyer, A., & Brown, R. A. (2017). Second-career science teachers' classroom conceptions of science and engineering practices examined through the lens of their professional histories. *International Journal of Science Education, 39*(11), 1511–1528.

Bang, E. J., & Luft, J. A. (2016). Practices and emerging identities of beginning science teachers in online and offline communities of practice. In L. Avraamidou (Ed.), *Studying science teacher identity: Theoretical, methodological and empirical explorations* (pp. 261–294). Rotterdam: Sense.

Beijaard, D., Verloop, N., & Vermunt, J. D. (2000). Teachers' perceptions of professional identity: An exploratory study from a personal knowledge perspective. *Teaching and Teacher Education, 16*(7), 749–764.

Bryce, N., Wilmes, S. E., & Bellino, M. (2016). Inquiry identity and science teacher professional development. *Cultural Studies of Science Education, 11*(2), 235–251.

Dawson, E. (2014). Equity in informal science education: Developing an access and equity framework for science museums and science centres. *Studies in Science Education, 50*, 209–247.

Dawson, E. (2017). Social justice and out-of-school science learning: Exploring equity in science television, science clubs and maker spaces. *Science Education, 101*, 539–547.

Dreon, O. (2008). *New science teachers' descriptions of inquiry enactment* (Doctoral dissertation). Retrieved from https://www.researchgate.net/profile/Oliver_Dreon/publication/253123692_New_science_teachers%27_descriptions_of_inquiry_enactment/links/571e06bc08aed056fa2261bc/New-science-teachers-descriptions-of-inquiry-enactment.pdf.

Dunlop, L., Turkenburg-van Diepen, M., Knox, K. J., & Bennett, J. (2020). Open-ended investigations in high school science: Teacher learning intentions, approaches and perspectives. *International Journal of Science Education, 42*(10), 1715–1738.

Eick, C. J., & Reed, C. J. (2002). What makes an inquiry-oriented science teacher? The influence of learning histories on student teacher role identity and practice. *Science Education, 86*(3), 401–416.

El Nagdi, M., Leammukda, F., & Roehrig, G. (2018). Developing identities of STEM teachers at emerging STEM schools. *International Journal of STEM Education, 5*(36). https://doi.org/10.1186/s40594-018-0136-1

Grabill, J. T. (1998). Utopic visions, the technopoor, and public access: Writing technologies in a community literacy program. *Computers and Composition, 15*, 297–315.

Haslam, S. A. (2017). The social identity approach to education and learning: Identification, ideation, interaction, influence and ideology. In K. I. Mavor,

M. J. Platow, & B. Bizumic (Eds.), *Self and social identity in educational contexts* (pp. 19–52). Oxford: Routledge.

Haslam, S. A., Reicher, S. D., & Platow, M. J. (2011). *The new psychology of leadership: Identity, influence and power.* New York: Psychology Press.

Irving-Bell, D. M. (2018). *The formation of science, technology, engineering and mathematics teacher identities: Pre-service teachers' perceptions* (Doctoral dissertation). Retrieved from https://eprints.lancs.ac.uk/id/eprint/127684/1/2018DawneMarilynIrving_BellPhD.pdf.

Luehmann, A. (2016). Practice-linked identity development in science teacher education: GET REAL! Science as a figured world. In L. Avraamidou (Ed.), *Studying science teacher identity: Theoretical, methodological and empirical explorations* (pp. 15–48). Rotterdam: Sense.

Luehmann, A. L., & Markowitz, D. (2007). Science teachers' perceived benefits of an out-of-school enrichment programme: Identity needs and university affordances. *International Journal of Science Education, 29*(9), 1133–1161.

Manning, A. (2017). *Urban science teachers exploring how their views and experiences can influence decisions to remain in post or not* (Doctoral dissertation). Retrieved from https://core.ac.uk/download/pdf/141244962.pdf.

McIntyre, J., & Hobson, A. J. (2016). Supporting beginner teacher identity development: External mentors and the third space. *Research Papers in Education, 31*(2), 133–158.

Melville, W., Bartley, A., & Fazio, X. (2013). Scaffolding the inquiry continuum and the constitution of identity. *International Journal of Science and Mathematics Education, 11*(5), 1255–1273.

Molander, B. O., & Hamza, K. (2018). Transformation of professional identities from scientist to teacher in a short-track science teacher education program. *Journal of Science Teacher Education, 29*(6), 504–526.

Nieswandt, M., Barrett, S. E., & McEneaney, E. H. (2013). Predictors of science subject discipline identities: A statistical analysis. *Canadian Journal of Science, Mathematics and Technology Education, 13*(1), 90–110.

Pantić, N., & Florian, L. (2015). Developing teachers as agents of inclusion and social justice. *Education Inquiry, 6*(3), 333–351.

Porter, J. E. (1998). *Rhetorical ethics and internet worked writing.* Greenwich, CT: Ablex.

Rawls, J. (1971). *A theory of justice.* Cambridge, MA: Harvard University Press.

Richmond, G. (2016). Making sense of the interplay between identity, agency and context in the development of beginning science teachers in high-poverty schools. In L. Avraamidou (Ed.), *Studying science teacher identity: Theoretical, methodological and empirical explorations* (pp. 219–238). Rotterdam: Sense.

Rivera Maulucci, M. S. (2013). Emotions and positional identity in becoming a social justice science teacher: Nicole's story. *Journal of Research in Science Teaching, 50*(4), 453–478.

Rushton, E. A. C., & Reiss, M. J. (2019). From science teacher to 'teacher scientist': Exploring the experiences of research-active science teachers in the UK. *International Journal of Science Education, 41*(11), 1541–1561.

Rushton, E. A. C., & Reiss, M. J. (2020). Middle and high school science teacher identity considered through the lens of the social identity approach: A systematic review of the literature. *Studies in Science Education*. https://doi.org/10.1080/03057267.2020.1799621

Snyder, C., Oliveira, A. W., & Paska, L. M. (2013). STEM career changers' transformation into science teachers. *Journal of Science Teacher Education, 24*(4), 617–644.

Ufnar, J. A., & Shepherd, V. L. (2019). The Scientist in the Classroom Partnership program: An innovative teacher professional development model. *Professional Development in Education, 45*(4), 642–658.

Varelas, M., House, R., & Wenzel, S. (2005). Beginning teachers immersed into science: Scientist and science teacher identities. *Science Education, 89*(3), 492–516.

Walkington, H., & Rushton, E. A. C. (2019). Ten salient practices for mentoring student research in schools: New opportunities for teacher professional development. *Higher Education Studies*. https://doi.org/10.5539/hes.v9n4p133

Woolhouse, C., & Cochrane, M. (2015). Educational policy or practice? Traversing the conceptual divide between subject knowledge, pedagogy and teacher identity in England. *European Journal of Teacher Education, 38*(1), 87–101.

Young, I. M. (1990). *Justice and the politics of difference*. Princeton, NJ: Princeton University Press.

10

Developing Professional Practice as a Teacher Scientist

10.1 Introduction

As has been explored in the previous chapter, I suggest that teachers who work with students in research develop a distinct professional identity, namely, the teacher scientist. This is a theoretical model, elicited from a detailed analysis of the experiences of 53 teachers and technicians who have worked in this way with their school students. In this final chapter, I share a framework of practices for teachers and technicians who support school student independent research. In sharing these practices, I aim to provide teachers and technicians with framework against which to reflect upon their current practice and, also, to share ways of working that may further extend this aspect of their work. Research that has considered the pedagogy of supporting and mentoring undergraduate research has identified effective practice that applies across disciplines and is not limited to science or STEM. In this chapter, the findings are grounded in the experience of high school science teachers and technicians; however, I argue that as with the undergraduate context, these practices have relevance to a broad range of disciplines including the humanities and the creative arts.

E. A. C. Rushton, *Science Education and Teacher Professional Development*, Palgrave Studies in Alternative Education, https://doi.org/10.1007/978-3-030-64107-8_10

10.2 Supporting Student Research: Insights from the Higher Education Context

The publication of ten salient practices for undergraduate research mentoring, based on a large-scale literature review (Shanahan, Ackley-Holbrook, Hall, Stewart, & Walkington, 2015), revealed an effective pedagogy of mentored research which applies across disciplines, institution types and countries. Subsequent research with award-winning undergraduate research mentors has begun to elicit nuances to this pedagogy, guided by a common set of underlying values that mentors hold (Walkington, Stewart, Hall, Ackley, & Shanahan, 2019): these manifest as an ability to establish and sustain a sense of challenge, while maintaining meaningful engagement and a sense of achievement amongst students. This research has revealed how award-winning mentors see the future of mentored research, particularly their commitment to broadening participation amongst a greater student demographic, and democratisation of the research process (Shanahan, Walkington, Ackley, Hall, & Stewart, 2017). Furthermore, undergraduate research mentoring has been conceptualised as a professional development activity when a co-mentoring model has been adopted (Ketcham, Hall, Fitzgibbon, & Walkington, 2018). The positive and negative impacts of undergraduate student research mentoring activity have also been explored in relation to career and identity development for university academics (Hall, Walkington, Shanahan, Ackley, & Stewart, 2018). There have been accounts of how mentoring practice varies across different countries (Larson, Partridge, Walkington, Wuetherick, & Moore, 2018), as well as utilisation of the ten salient practices to consider curriculum redesign to ensure a high-quality mentored experience where students receive research mentoring outside their own higher education institution (Hall et al., 2018).

This chapter shares ten salient practices for teachers mentoring high school student researchers (Walkington & Rushton, 2019). The practices are described in a sequence which attempts to mirror the research process that has been observed in the work of teachers and technicians who work as teacher scientists. This process frequently begins with the planning and scaffolding required to initiate research projects and ending with the research dissemination and publication phase. The practices presented here continue the work of Walkington and Rushton (2019), who explored the experiences of high school teachers who supported school students when undertaking independent research. Walkington and Rushton (2019) considered the differences between school and university education, where levels of expectation for students to engage in primary research differ. For example, teachers are in the main, partnering with students in research, which is new to both partners, a situation different to most undergraduate settings, where most university academics are likely to have knowledge and expertise in the particular area of research. However, teachers in schools may engage with multi-generational projects, becoming increasingly familiar with a research area where school student cohorts are 'passing through'. Yet there are also clear similarities between undergraduate and high-school research, in that many students take part in the research experiences voluntarily, that 'authentic' research (i.e. making a genuine contribution to knowledge) can be a shared experience between a teacher and students(s), and that students can take on greater control and ownership throughout the research process.

Thiry and Laursen (2011) identified three forms of support which undergraduate research mentors provide for student researchers in bringing them into a scientific community of practice: intellectual support (with the research process); personal and emotional support (being available to students and taking an interest); and support with professional socialisation (over and above disciplinary knowledge and skills, passing on the values and norms of scientific endeavour). While teacher and technician support for research projects in high schools will be highly likely to involve intellectual support, and in isolation could be termed 'supervision' rather than mentoring, the personal and emotional or 'whole student' support and professional socialisation as a scientist are perhaps less frequently anticipated forms of practice at school level. However, engaging in school-based mentoring as described in this article clearly moves from a simple teacher-student relationship to research mentoring and in some cases adopts a co-researcher model. In sharing the ten practices enacted by teachers, there are examples which speak to all three forms of Thiry and Laursen's (2011) support given by teachers and technicians. In what follows, I outline the ten salient practices of teachers and technicians who in supporting high school student research are working as teacher scientists.

10.3 Ten Salient Practices for Developing as a Teacher Scientist

1. *Do strategic pre-planning to be ready to respond to students' varying needs and abilities throughout the research process.*

It is challenging for teachers and technicians to anticipate pre-planning aspects of authentic research where there is a lack of familiarity with a specific research area and a genuine sense of 'stepping into the unknown' with students. However, school students represent a range of abilities and require strong support to access the knowledge, skills and concepts needed to engage actively at the outset of new research. This might be because they have not yet encountered material in the high school curriculum or because the research project extends into areas beyond the curriculum. In addition, teachers of high school science, while having a good knowledge of a broad science curriculum, would be unusual in having experience of research in an area that they are now actively investigating with their students. One form of pre-planning is appropriate project selection and student recruitment. In the main, teachers and technicians choose the research projects and lead on student recruitment. Recruitment is predominantly based upon identifying student interest through open invitations, rather than proven academic ability, but can also be constrained by structural issues, such as timetabling. Students need to be available at the same time so that teachers can provide access to the research projects, for example, after school or during lunchtime. However, some teachers do undertake pre-planning, for example pre-teaching scientific content or some of the core knowledge prior to starting the research project, so that students have the necessary background to begin the research. Another aspect of pre-planning is working to timelines, where teachers try to align what can be a messy or jerky research journey to appropriate school timescales. Timelines for the research may be set externally where students are working to external qualification frameworks such as the Extended Project Qualification (EPQ) and CREST awards[2]. In this case, the role of teacher is to help students stay focused and on schedule, although this is something that teachers find students take increasing responsibility for as the research progresses. Although teachers and technicians are often learning alongside their students, they encourage a positive attitude that with time and preparation, learning will be effective, and the project outcomes will be met.

2. *Set clear and well-scaffolded expectations for student researchers.*

Over time, teachers and technicians who support high school student research tend to set clear expectations and provide scaffolding. This strong support is often provided through setting homework tasks as prompts or starter activities for research to develop student interest and understanding of the key concepts before they undertake the main research phase. Teachers and technicians frequently break down research into more manageable tasks for students. When working with students to prepare for a conference presentation, for example, teachers guide students by helping them divide the content into sections for individual students to prepare. Teachers and technicians also provide opportunities for students to rehearse presentations to audiences of student groups and other teachers. Some teachers also prepare students who present at conferences by describing the environment and audience in a way that highlights to the students the professional nature of the event so that they understood the expectations their teacher and technicians have of them, including implicit expectations about how to present and to operate effectively in a new environment. Teachers provide examples of previous work (e.g. high-quality research posters) to students individually as well as displaying them in their school departments to give students a framework and to set high expectations. Teachers also use explicit frameworks to give students additional guidance and support in how to undertake scientific research and to share their expectations of the student from the outset, for example, the expectation that the student will mentor younger students. Some teachers and technicians suggest that rules and contracts are not part of their ‘informal’, ‘after-school’ approach to research projects and that their research with students is founded upon trust and student motivation. When teachers do employ agreements, they often link them closely to the curriculum guidelines associated with external frameworks, for example, the course specification for the EPQ. In this way, the teacher integrated their practice of setting clear expectations (practice two) with providing opportunities for student researchers to be involved with peer and near-peer mentoring (see practice nine below).

3. *Guide students in the technical skills, methods and techniques of conducting research in the discipline.*

A starting point for teachers and technicians when encouraging students to explore and experience science research is the introduction of Journal Clubs. These sessions are often held at lunchtimes or during after-school sessions where students, usually aged 14–18 years, read and discuss articles with their teachers. Journal Clubs are used to meet several objectives simultaneously: firstly, to understand current research; secondly, to consider links between this current research and topics in the school science curriculum; thirdly, to understand the scientific method through how a journal article is structured from the wider context and the need for the research to results, discussion and next steps and fourthly, mentors report using the further questions offered at the end of articles to position their students as being an active part of the scientific community where they can continue the research to try to answer those questions or others that they may have. Teachers and technicians that implement Journal Clubs also describe the progression of these activities from ones that are initially teacher-led, with teachers selecting the articles and providing verbal and/or written framework for discussion, to ones that are led by students, discussing the research that links to their own research project and interests. In this way, the teachers and technicians are modelling how to use wider research to inform new research endeavours and initiatives. This progression from teacher to student-led sessions is also a way to model salient practice seven, where student ownership of the research increases over time.

As part of guiding students in this practice, teachers and technicians frequently explain to their students how scientific research and inquiry works, emphasising that they (the teacher/technician) do not have the answers and supporting students to develop the thinking skills required to progress through different phases of research. Furthermore, teachers and technicians build a rapport with their students when they guide them through some technical aspects by sharing that they themselves are learning something new and are experiencing anxiety when encountering something new and unfamiliar. Some teachers and technicians do not teach the technical skills themselves but provide guidance to students as they learn skills from other partners they work with, for example, undergraduate students. Teachers use practical experiment time in the curriculum to build a scaffold for students so that they have the confidence to participate in the elective, non-compulsory research activities. Teachers also leverage the high-tech nature of some of the research equipment to encourage students to participate in research. Teachers and technicians highlight the ethical areas that need to be considered in research, both through discussions of published research and also through the planning and experimental phases of research projects which may include human participants or require field-based data collection.

4. *Balance rigorous expectations with emotional support and appropriate personal interest.*

Teachers and technicians emphasise that all aspects of scientific endeavour are part of a rigorous process and that students are going to have to become comfortable with not knowing an answer and working out what an answer might be from the information they have currently. The purpose here is to support students as they make a transition from learning through a transmission model from their teacher, to learning as a co-inquirer, something that prepares students for post-school education and employment. Another example of this transition is to move students from being positioned as learners to that of researchers through dissemination activities. Teachers and technicians encourage students to contribute to conferences in a way that engages the conference audience in the students' research. This means giving students explicit feedback about whether their poster submissions are appropriately and concisely sharing their research findings. Teachers also support students in making the most of the conference experience, encouraging them to ask questions of other student researchers and at times brokering that engagement by introducing students to other groups and initiating discussions. Teachers and technicians set high expectations for students by explaining to them at the outset that if they give their best, they have the capacity to make a valuable contribution, for example, publishing their work in a peer-reviewed journal.

Where they have the relevant prior experience, teachers and technicians take the time to support students through the rigorous process of writing a paper for publication by acknowledging both the challenges and the opportunities that this experience will bring for students. Where they do not have sufficient experience of writing papers, teachers and technicians broker relationships with others, for example, post-graduate researchers and other academics to support their school students. A key part of this process is achieved by explaining to students that what they are experiencing when 'writing up' their research reflects the reality of science and that they are part of a global science community. Teachers provide support for students when they persevere through the challenges presented by publication and perhaps initially struggle with the high level of expectations and attention to detail required by publication.

In addition, teachers and technicians recognise that students may need emotional support in order to excel as researchers, for example, when they are invited to share their research with the wider public as part of events that are external to their school community. Opportunities to engage the public with their research clearly comes from rigorous expectations, but in order for the students to be able to achieve this, they need the support of their teachers and technicians in terms of practising, confidence building, structuring the day and managing time and sharing different roles amongst students.

5. *Build a sense of community among the members of the research team.*

Teachers and technicians recognise that through research they, and their students, have the opportunity and capacity to be a contributing part of a wider research community that is beneficial to all partners. Teachers leverage a network of researchers across different institutions in a range of roles including undergraduate students, academics, technical support and public engagement staff. Teachers and technicians are skilled at creating spaces beyond the official timetable to develop a research team and community, for example, through establishing a journal or after-school club and frequently include peer and near-peer mentoring as part of high school research teams. Some of school-based research projects evolve over a number of years, involving multiple generations of students who are motivated to work together, those with greater experience helping those with less, leading to 'near peer' or 'vertical' mentoring communities. For long-standing projects implemented over a number of years, teachers and technicians active recruit project alumni (e.g. undergraduate, doctoral and post-doctoral researchers) to mentor current school students, for example, writing peer-reviewed papers using data from projects that they themselves contributed to whilst at school.

As well as building a sense of community through peer and near-peer mentoring, teachers and technicians also organise research-related trips (e.g. a visit to their local university, science exhibitions) and organising external speakers to visit the school (e.g. STEM Ambassadors, local conservation groups, university speakers). All these activities further engender the notion that research is a group endeavour as part of a wider scientific community. Teachers and technicians often collaborate with other schools and use journal clubs or after-school clubs to develop a sense of community within their schools and local areas. Compared to the experiences of support for student research in the higher education context, school teachers and technicians very occasionally describe social events, for example, meeting for lunch in a local restaurant as part of an end of term or project celebration event or incorporating a social aspect to a research focused event, such as having a coffee in a local café as part of an external visit to share student research with the general public. Many teachers suggest that they would like to include a social aspect in the future as they recognise the value of these opportunities for students, but that they are limited by a range of factors including financial, staffing and time constraints.

6. *Dedicate time to one-on-one, hands-on mentoring.*

Compared to the higher education context, teachers and technicians are less able to provide one-to-one, hands-on mentoring as this is an infrequent part of their day-to-day work, where the vast majority of work with students occurs in groups. When teachers and technicians do use one-to-one mentoring, this is predominantly used to encourage students to persevere during challenging phases of research, coaching individuals to go beyond their own expectations of their ability. Teachers provide additional support for students who are anxious about particular phases of research, for example, extra rehearsals with shy students preparing to present at conferences. Technicians describe that they can provide an 'open door policy' during times when students are free (e.g. lunchtime sessions, study periods) which enable them to 'troubleshoot' problems (often with software, equipment) due to the relatively greater flexibility in their working day compared with teachers.

7. *Increase student ownership of the research over time.*

As indicated in salient practice two, teachers have the expectation that student ownership of research will increase over time. Teachers and technicians direct the research at the outset, prompting students through questions and identifying tasks that need completing. Over time, some students take on that role and direct their peers. Ownership of the management and progression of the research can be taken on by some students organically and in other contexts teachers hand over parts of the research project to nominated students, for example presentations to external audiences, which fosters student autonomy and their ownership of project outputs.

8. *Support students' professional development through networking and explaining the norms of the discipline.*

Providing development opportunities for school-based researchers includes involvement in networking opportunities, preparing to share research findings (e.g. co-producing research papers, posters or other forms of research dissemination) and helping students to access work experience or placement opportunities. The most common means to develop networking is by introducing students to researchers, whether at school/college through invited talks or in-school sessions with visiting researchers, or externally through site visits. Informal networking can be particularly powerful for students, especially where pupils can identify role models, such as university students who may be easier to identify with than more eminent scientists. Teachers who have been involved in research with their students for some time are able to create opportunities for their students to share their research (e.g. presenting to primary school groups), and through that build networks with other researchers and engage younger students in scientific endeavour through their experiences. Through creating these opportunities, teachers are explaining to their students the norms of the discipline and the social/professional norms of an academic environment. Teachers and technicians facilitate networking for students, co-ordinating sessions led by external speakers and intentionally stepping back, allowing their students to engage and build a professional relationship with external speakers and providing follow-up activities for students to make best use of an external speaker's time. Teachers encourage students to share their research at events where they meet scientists and can discuss science as contributing members of the research community. School students require support from teachers and technicians to make best use of networking opportunities at events, with teachers facilitating conversations between school students and PhD students. Brokering opportunities to allow students to ask questions and understand the career paths and experiences of early career researchers that they can identify with is a key element of this practice. Teachers and technicians frequently support student development, and some achieve this by helping students obtain work experience and research placements that vary in length (a few days to four weeks) and context (industry, university/research institute laboratory, hospital). External opportunities feature in student professional development and, in addition to work experience and research placements, include supporting students to connect with external scientists, visit specialist libraries, co-publish research papers and present at external conferences. Teachers and technicians provide opportunities for student professional development within the school context, for example, working with other schools, presenting at school conferences and displaying students' work in school.

9. *Create opportunities for peer and near-peers to learn mentoring skills and give more students access to research opportunities.*

Teachers show students that being a successful researcher is being in an intellectual space where answers are not yet known which is a space that they do not frequently encounter during their experience of high school. Modelling by teachers and technicians of researcher characteristics, such as being comfortable with not knowing the answer, opens up a space for all participants to become experts. It allows mentoring between peers and near-peers to become part of this space, even when younger school students are mentoring older ones, having developed particular expertise. As school research is predominantly a team activity, teachers frequently incorporate peer mentoring and near-peer mentoring into the delivery of their research activities. The role of teachers in this situation is in ensuring that peer and near-peer mentoring is productive and respectful. Some teachers and technicians establish rules for behaviour and expectations and involving peers or near-peers in mentoring as part of teachers' practice. These near-peers included older school students, alumni, university students, teachers with research experience and students with research project experience. Where possible, teachers and technicians also involve partner schools, STEM ambassadors and university academics. Some teachers and technicians include vertical mentoring as an integral part of the research project whereas other teachers do not delegate mentoring because they identified it as being a core part of their role.

10. *Encourage students to share their findings and provide guidance on how to do so effectively in presentations and in writing.*

Dissemination of research is actively supported by teachers and technicians as a fundamental part of the research process. Examples of this practice include providing templates for posters, running sessions to discuss what makes a good poster, as well as supporting the preparation and rehearsal of presentations. Raising awareness of competitions and avenues outside school for students to share their work, for example, talks to local societies and visits to other schools, were opportunities for dissemination to have greater authenticity. Examples of dissemination within the school community included school assemblies, subject societies and science fairs, as well as presentations to visiting schools and parents, with more long-term displays of posters in the school. External dissemination was concentrated on large-scale events such as the Royal Society Science EXPO, partner university conferences and the Big Bang Fair. Teachers and technicians regularly take their students to conferences. For conference events, where the emphasis is on oral presentations of research, perhaps with poster presentations, teachers distinguish the support that they gave to their research mentees into activities done before, during and after the conference event. Preparing posters and practising presentations was important, with audiences for these practice sessions including parents, other teachers and other students. For example, an English teacher was invited to coach students in presentation skills and how to build a narrative and divide a presentation up into tasks for students, then help them put it back together as a group presentation. Explaining expectations is also important and includes time to discuss the benefits of collaboration and communication in science. Practise sessions are also used to encourage students that their work is a valuable and high-quality contribution. During conference events, the role played by teachers and technicians shifts to a supporting one so that students feel comfortable in the unfamiliar environment and are able to make full use of the opportunity. Introducing students to other people and facilitating conversations to develop networks is an important role for teachers and technicians. Sometimes this is focused on developing networks with other schools, but also to network with eminent speakers. Enabling students to be independent and 'step up to the challenge' was also important for teachers and technicians. Following conference events, work such as raising awareness through the school community is important, to maximise the legacy from the event and share it with others. Teachers and technicians who are more experienced in supporting high school student research ensure that after the conference there are opportunities for students to discuss the way that knowledge gained at the conference shapes the future direction of their research, so that presentation was not seen as an endpoint, but more like a critical event or stimulus for the project. While research is very often presented orally or through posters and displays, publication in scientific journals is a less frequent outcome for school student research projects. Teachers and technicians with experience of supporting students in writing peer-reviewed articles describe their role as liaising with academics who are more familiar with this form of writing rather than actually supporting the writing themselves.

In summary, although teachers and technicians generally have less experience of disseminating research, they play an important role in identifying dissemination opportunities and brokering relationships with external mentors who can enable students to publish their research in peer-reviewed journals and present at conferences. In supporting students to develop working relationships with academic mentors, teachers model effective research collaborations, where individuals have different roles with complementary areas of expertise.

10.4 The Ten Salient Practices as a Tool for Professional Reflection

Although the ten salient practices have been set out as distinct or separate practices, Walkington and Rushton (2019) argue that together they form a holistic pedagogy. Many of the activities which teachers and technicians engage in to support their high school students in authentic research feature multiple practices, simultaneously. The school context and the nature of the selected research project can impact on the pedagogic approach adopted, as can a teacher's and technician's own prior experience and interests. As Walkington and Rushton (2019) have previously highlighted the pedagogy of supporting student research is broadly consistent between undergraduate and school settings, however, there was a difference in emphasis. As might be expected, teachers and technicians provided much greater support and framing for high school students (practice two) than academic mentors in higher education. Teachers and technicians also provided much more guidance and modelling for students when networking (practice eight). Creating a sense of community (practice five) was well developed by teachers as they tend to work with groups on projects. Interestingly this sense of research community is something that academics in university and college settings found the most challenging to develop (Walkington, Hall, Shanahan, Ackley, & Stewart, 2018) outside of laboratory settings. As teachers are working with minors, there are limits to the types of one-to-one (practice six) and informal socialising (practice five) that might be developed in schools, which contrasts to networking norms in higher education research contexts. Identifying teachers' practices in this way is consistent with research that has identified six qualitatively different teacher perceptions of implementing open-ended investigations with high school students of post-16 science (Dunlop, Turkenburg-van Diepen, Knox, & Bennett, 2020). By sharing the six perceptions and intended learning outcomes of teachers (as outlined in Chap. 9), Dunlop et al. (2020) suggest that they provide a framework for teacher reflection as part of professional development that could both further develop the practice of teachers who already work in this way or enable teachers seeking to implement open-ended investigations for the first time. Consistent with Dunlop et al. (2020), by outlining ten salient practices of teachers who are research active with their students, my intention is to create a useful narrative or framework that will support teachers and technicians to reflect upon their current practice or to develop new ways of working with students such that further research can be supported in schools across many disciplines. By reflecting upon the ten salient practices in the context of the model of teacher scientist identity outlined in Chap. 9, those engaged or seeking to support school-based student research can actively reflect upon the ways in which their practice could develop different facets of their professional identity. For example, providing personalised support for students depending on their age, motivation and skills (practice one) and providing emotional support (practice four) are ways that a teacher can develop their social justice identity through enabling participation in research for the greatest diversity of students. Through giving students guidance as to the skills, methods and techniques of a discipline (practice three) and explaining the norms of the discipline (practice eight), teachers and technicians can develop their subject identity by actively engaging in the research processes of that subject. Finally, in setting well-scaffolded expectations for students (practice two), building a sense of community (practice five) and in enabling increased student ownership of the research overtime (practice seven), teachers and technicians are engaged in the work of inquiry. Further research is needed to explore the ways in which these practices can be integrated into the teacher scientist model of identity across a range of contexts and disciplinary areas; however, I argue that together these practices and model of identity form a productive and purposeful way of exploring practice at an individual level.

In the final section, I draw together the key findings of this research but first, I consider the implications and future directions of this work.

10.5 Future Directions and Implications

To date, the teacher scientist model of research-active teachers and technicians has been drawn from science and science-related subjects; future empirical research in this area will further our understanding of the specific ways in which this approach to teacher professional development could extend to other subjects and involve schools and colleges across the sector. As documented by Shanahan et al. (2015), undergraduate research as an approach to teaching has broadened participation over the last decade to include (1) research that extends beyond the sciences to include social sciences, humanities, arts and professional studies, (2) institutions across the undergraduate sector and (3) alternatives to the traditional apprenticeship-model with an academic in a supervisory role, including course-based approaches that are arguably more democratic and reach a more academically and socio-economically diverse group of students. This suggests that broader participation in school settings is indeed possible beyond STEM subjects. Further research might fruitfully consider whether being research active is a productive way for teachers of subjects including those from the creative and performing arts, humanities and social sciences may develop and sustain their subject-specialist identities.

Future work might also systematically consider how this approach to teaching has been and could be a more widely available opportunity for school-based educators and their students. This could include (1) how research forms part of school curricular and extra-curricular offerings, (2) how teachers are supported by Subject Associations and Learned Societies to be research active and (3) how this approach to professional development is integrated with Initial Teacher Training and Education. Over the course of this current research, I have highlighted a number of curricular (A-level, Extended Project Qualification, International Baccalaureate) and extra-curricular (CREST Awards) pathways for teachers to be research active in schools, and this has been a particular focus for research that considers science education in high school settings (Dunlop et al., 2020; Moote, Williams, & Sproule, 2013; Rushton & Reiss, 2019; Rushton et al., 2019). More work is needed to understand the affordances and limitations in this area so that these opportunities are available to as many students and teachers as possible. The support of Subject Associations and Learned Societies has also been positively noted by teachers and technicians who have engaged with initiatives such as the Royal Society Partnership Grant. A comprehensive study which considers the support of and collaboration with school teachers and their students in the context of research would further elucidate the impact of these initiatives. Lastly, when considering the role of teacher education programmes in the context of research-active teachers, it is important to recognise the pressures that these institutions face in terms of delivering courses that strike a balance between pedagogical instruction, subject knowledge enhancement and classroom-based practice. And yet, there is wider recognition (e.g. Grier & Johnston, 2009; Molander & Hamza, 2018; Snyder, Oliveira, & Paska, 2013) of the critical role that teacher education programmes have in providing all teachers with the opportunity to explore and develop their professional identity. With that in mind, I suggest that the model of being actively engaged in subject-specific research espoused by the teacher scientist could be incorporated into the current content of these programmes, for example through existing sessions that consider identity and professional development, rather than being a further item to add to already busy programmes. This approach is consistent with that outlined in the Initial Teacher Training Core Content Framework (DfE, 2019) which highlights the need for teachers to 'strengthen pedagogical and subject knowledge by participating in wider networks' and, 'engage critically with research' (p. 29).

Finally, a subsequent phase of research might consider the ways in which the teacher scientist model might support the retention of teachers in the profession. Although I do not propose that the teacher scientist approach to teacher professional development as a panacea, or an alternative to providing socially just professional contexts (e.g. fair pay, appropriate working conditions and mutual trust and respect), I do contend that further theoretical understanding and empirical validation of the model in other disciplines would make a similarly useful contribution to the recent focus on teacher professionalisation through engagement in pedagogical research, for example through the work of the Chartered College of Teaching and The Research Schools Network (a collaboration between the Education Endowment Foundation and the Institute for Effective Education). Currently, many initiatives are focused on ameliorating the global crisis in science teacher recruitment and retention, but the challenges extend beyond this particular group of teachers, and I suggest that the teacher scientist model has a wider relevance for the field of education than that which solely concerns science teachers.

10.6 Conclusion

Over the course of this book, I have considered in detail the experiences of science teachers and technicians working in high schools who are research active with their students. These research projects have been subject-focused, rather than pedagogical in nature, and have been largely situated in the contexts of biology, chemistry and physics. There have been variations in the terms of the research foci, the project duration, the extent of adult versus student leadership, the provision of funding and external support and the place (or not) of the research projects within the school curriculum. The impetus for this research came from the observation that although much attention has been given to the experiences of students who participate in independent research projects (Bennett, Dunlop, Knox, Reiss, & Torrance Jenkins, 2018; Minner, Levy, & Century, 2010), far less attention has been given the experiences of teachers and technicians, save for a few notable exceptions (Dunlop et al., 2020; Rushton & Reiss, 2019). My purpose in highlighting the five empirical themes of (1) Freedom to teach, (2) (Re)connection with science/research, (3) Collaboration, (4), Professional development and (5) Student/societal development through research has been to document the experiences of over 50 teachers and technicians so to better understand this area of their professional practice and how this can and does shape their professional identity. This empirical work has resulted in the proposition of a model of identity development with three facets, namely subject identity, inquiry identity and social justice identity. When considering these facets in the context of the work of other researchers in the field, including both studies from England (e.g. Dunlop et al. (2020)) and the USA (e.g. Ufnar and Shepherd (2019)), I have noted commonalities in teachers' focus on independent research projects to support and enhance student learning as well as enabling their own professional collaborations and flourishing. I have argued that through the mechanisms of identification, ideation, interaction, influence and ideology, rooted in the Social Identity Approach (SIA), individuals are able to develop and sustain their identity as a teacher scientist, and that this focus on group-membership provides an important source for teachers' positive professional identity. Finally, I have shared a framework of ten salient practices (Walkington & Rushton, 2019) to provide both a source of reflection and guidance for those who wish to further develop their professional development in this way. This work provides a substantial insight into the lived experiences of teachers and technicians who work with school students in this way. In order to capture the broadest perspective of this approach to professional development, I have outlined areas for future research which include:

1. teachers' experience from contexts which extend beyond the sciences,
2. systematically examining the implementation and nature of this approach and the different types of support available, for example, through Subject Associations and Learned Societies and
3. the extent to which this model of teacher professional development aids retention.

Such future directions for research should include large-scale, longitudinal studies as well as detailed narrative case studies, including life history, autobiographical approaches—and greater co-authorship with teachers.

This has the potential to ensure that teacher professional development guidance generated from research which is relevant to the fullest range of teachers and contexts.

In bringing this research to a conclusion, I wanted to briefly return to the reflections that teachers and technicians shared when considering *why* they sought out and persevered through challenges to provide opportunities for their students to participate in research projects as part of their science education. Teachers and technicians frequently described the potential for school students to experience science education as the joy of discovering new ideas about their world and the opportunity to share those discoveries with others, as part of a wider scientific community. As Natalie (chemistry teacher) said, 'Science is about the freedom of the unknown and of discovery and research projects help students understand that'. Throughout this book, I have argued that the freedom and discovery that this approach brings for students is also a rich opportunity for the professional development of teachers and technicians that enables them not only to enact the science curriculum but also to contribute to science itself.

References

Bennett, J., Dunlop, L., Knox, K. J., Reiss, M. J., & Torrance Jenkins, R. (2018). Practical independent research projects in science: A synthesis and evaluation of the evidence of impact on high school students. *International Journal of Science Education, 40*(14), 1755–1773.

Department for Education. (2019). *Initial teaching training: Core content framework*. Retrieved June 20, 2020, from https://www.gov.uk/government/publications/initial-teacher-training-itt-core-content-framework.

Dunlop, L., Turkenburg-van Diepen, M., Knox, K. J., & Bennett, J. (2020). Open-ended investigations in high school science: Teacher learning intentions, approaches and perspectives. *International Journal of Science Education, 42*(10), 1715–1738.

Grier, J. M., & Johnston, C. C. (2009). An inquiry into the development of teacher identities in STEM career changers. *Journal of Science Teacher Education, 20*(1), 57–75.

Hall, E., Walkington, H., Shanahan, J., Ackley, E., & Stewart, K. (2018). Mentor perspectives on the place of undergraduate research mentoring in academic identity and career development: An analysis of award-winning mentors. *International Journal for Academic Development, 23*(1), 15–27. https://doi.org/10.1080/1360144X.2017.1412972

Hall, E., Walkington, H., Vandermaas-Peeler, M., Shanahan, J. O., Kolbech-Gudiksen, R., & Mackenzie-Zimmer, M. (2018). Enhancing short-term undergraduate research experiences in study abroad: Curriculum design and mentor development. *Perspectives on Undergraduate Research Mentoring 7.1* [online]. Retrieved from https://blogs.elon.edu/purm/2018/10/22/enhancing-short-term-undergraduate-research-experiences-in-study-abroad-curriculum-design-and-mentor-development-purm-7-1/

Ketcham, C., Hall, E., Fitzgibbon, H., & Walkington, H. (2018). Co-mentoring in undergraduate research: A faculty development perspective. In M. V. Peeler, P. Miller, & J. Moore (Eds.), *Excellence in mentoring undergraduate research* (pp. 155–179). Washington, DC: Council on Undergraduate Research.

Larson, S., Partridge, L., Walkington, H., Wuetherick, B., & Moore, J. (2018). An international conversation about mentored undergraduate research and inquiry and academic development. *International Journal for Academic Development, 23*(1), 6–14. Retrieved from http://www.tandfonline.com/doi/full/10.1080/1360144X.2018.1415033

Minner, D. D., Levy, A. J., & Century, J. (2010). Inquiry-based science instruction—What is it and does it matter? Results from a research synthesis years 1984 to 2002. *Journal of Research in Science Teaching, 47*(4), 474–496.

Molander, B. O., & Hamza, K. (2018). Transformation of professional identities from scientist to teacher in a short-track science teacher education program. *Journal of Science Teacher Education, 29*(6), 504–526.

Moote, J. K., Williams, J. M., & Sproule, J. (2013). When students take control: Investigating the impact of the CREST inquiry-based learning program on self-regulated processes and related motivations in young science students. *Journal of Cognitive Education and Psychology, 12*(2), 178–196.

Rushton, E. A. C., Charters, L., & Reiss, M. J. (2019). The experiences of active participation in academic conferences for high school science students. *Research in Science and Technological Education*, https://doi.org/10.1080/02635143.2019.1657395.

Rushton, E. A. C., & Reiss, M. J. (2019). From science teacher to 'teacher scientist': Exploring the experiences of research-active science teachers in the UK. *International Journal of Science Education, 41*(11), 1541–1561.

Shanahan, J., Ackley-Holbrook, E., Hall, E., Stewart, K., & Walkington, H. (2015). Salient practices of undergraduate research mentors: A review of the literature. *Mentoring and Tutoring, 23*(5), 359–376. https://doi.org/10.1080/13611267.2015.1126162

Shanahan, J. O., Walkington, H., Ackley, E., Hall, E., & Stewart, K. (2017). Award winning mentors see democratization as the future of undergraduate research. *Council on Undergraduate Research Quarterly, 37*(4), 4–11. https://doi.org/10.18833/curq/37/4/14

Snyder, C., Oliveira, A. W., & Paska, L. M. (2013). STEM career changers' transformation into science teachers. *Journal of Science Teacher Education, 24*(4), 617–644.

Thiry, H., & Laursen, S. L. (2011). The role of student–advisor interactions in apprenticing undergraduate researchers into a scientific community of practice. *Journal of Science Education and Technology, 20*, 771–784. https://doi.org/10.1080/00221546.2011.11777209

Ufnar, J. A., & Shepherd, V. L. (2019). The Scientist in the Classroom Partnership program: An innovative teacher professional development model. *Professional Development in Education, 45*(4), 642–658.

Walkington, H., Hall, E., Shanahan, J., Ackley, E., & Stewart, K. (2018). Striving for excellence in mentoring undergraduate research: The challenges and approaches to ten salient practices. In M. V. Peeler, P. Miller, & J. Moore (Eds.), *Excellence in mentoring undergraduate research* (pp. 105–125). Washington, DC: Council on Undergraduate Research.

Walkington, H., & Rushton, E. A. C. (2019). Ten salient practices for mentoring student research in schools: New opportunities for teacher professional development. *Higher Education Studies*. https://doi.org/10.5539/hes.v9n4p133

Walkington, H., Stewart, K. A., Hall, E. E., Ackley, E., & Shanahan, J. O. (2019). Salient practices of award-winning undergraduate research mentors—Balancing freedom and control to achieve excellence. *Studies in Higher Education*, 1–14. https://doi.org/10.1080/03075079.2019.1637838

Appendix

List of key informants quoted in Chaps. 4, 5, 6, 7 and 8

Amy, physics teacher
Annie, general science teacher
Anthony, physics teacher
Arthur, physics teacher
Bailey, technician
Barbara, chemistry teacher
Bella, physics teacher
Bethany, physics teacher
Clare, chemistry teacher
Dean, biology teacher
Declan, chemistry teacher
Dominic, physics teacher
Ellen, general science teacher
Elliott, physics teacher
Francis, physics teacher
Gordon, physics teacher
Hasan, physics teacher
Jacey, psychology teacher
Jacinta, physics teacher
James, physics teacher
Jane, biology teacher
Joan, technician

E. A. C. Rushton, *Science Education and Teacher Professional Development*, Palgrave Studies in Alternative Education, https://doi.org/10.1007/978-3-030-64107-8

John, physics teacher
Jonny, chemistry teacher
Joseph, physics teacher
Joshua, physics teacher
Keith, chemistry teacher
Mabel, biology teacher
Madeleine, general science teacher
Marina, biology teacher
Marion, biology teacher
Mark, physics teacher
Melanie, technician
Michelle, physics teacher
Natalie, chemistry teacher
Nathan, biology teacher
Ned, chemistry teacher
Peter, biology teacher
Rabail, technician
Ralph, physics teacher
Robert, chemistry teacher
Rose physics teacher
Sally, biology teacher
Sarah, technician
Sophie, biology teacher
Stephen, psychology teacher
Tahira, physics teacher
Tina, general science teacher
Tony, biology teacher

Index[1]

[1] Note: Page numbers followed by 'n' refer to notes.

E. A. C. Rushton, *Science Education and Teacher Professional Development*, Palgrave Studies in Alternative Education, https://doi.org/10.1007/978-3-030-64107-8

U

W

The manufacturer's authorised representative in the EU is Springer Nature Customer Service Centre GmbH, Europaplatz 3, 69115 Heidelberg, Germany. If you have any concerns regarding our products, please contact ProductSafety@springernature.com

Printed and bound by CPI Group (UK) Ltd, Croydon, CR0 4YY
05/10/2026
02227638-0001